W9-CDE-471

Exploring Chaos

Nina Hall was physical sciences editor of *New Scientist* at the time that she organized the series of articles in this collection. She is now working on the start-up of a new monthly science magazine in London. Ms Hall has a chemistry degree from the University of Oxford. After working in industry and having a family, she worked as features editor on *Chemistry in Britain,* and later as science correspondent on the *Times Higher Education Supplement.*

Exploring Chaos

A Guide to the New Science of Disorder

Edited by Nina Hall

W. W. NORTON & COMPANY
New York London

Copyright © 1991 by IPC Magazines New Scientist
First published in book form in England in 1992 by Penguin Books Ltd. under the title *The New Scientist Guide to Chaos.*

First American Edition 1993

All rights reserved

Printed in the United States of America
Manufacturing by Arcata Graphics/Halliday

Library of Congress Cataloging-in-Publication Data

Exploring chaos : a guide to the new science of disorder / edited by
Nina Hall.—1st American ed.
p. cm.
1. Chaotic behavior in systems. I. Hall, Nina.
Q172.5.C45E98 1993
003'.7—dc20
92-35082

ISBN 0-393-03440-2

W. W. Norton & Company, Inc., 500 Fifth Avenue, New York, N.Y. 10110
W. W. Norton & Company Ltd., 10 Coptic Street, London WC1A 1PU

1 2 3 4 5 6 7 8 9 0

Contents

Introduction

We are all aware of how small events can drastically change the course of history: the assassin's bullet that triggers a revolution, a chance meeting with a stranger at a party, the impulsive decision to catch a plane doomed to crash. Romantic novels are full of such stories. We expect life to be complicated and uncertain, scattered with random events that make the future difficult to predict. For this reason, many people find the 'pure' predictability of traditional science unattractive and difficult to relate to their own lives.

Recently, however, a new line of scientific inquiry called 'chaos theory' has caught the popular imagination. It seems to link our everyday experiences to the laws of nature by revealing, in an aesthetically pleasing way, the subtle relationships between simplicity and complexity and between orderliness and randomness.

Scientists have always searched for simple rules, or laws, that govern the Universe. For example, Isaac Newton could explain how the stars appeared to move across the sky with his simple laws of motion and theory of gravitation. At the beginning of the 19th century, the famous French mathematician Pierre Simon de Laplace believed firmly in a Newtonian universe that worked on clockwork principles. He proposed that if you knew the position and velocities of all the particles in the Universe, you could predict its future for all time.

This deterministic view received its first blow in the 1920s, when quantum mechanics was developed to describe the world of the very small. It explained how fundamental particles such as the electron behaved. But quantum mechanics is a statistical theory based on probability, and it is impossible to measure the position and momentum of a particle at the same time. This so-called uncertainty principle of quantum theory seems to be inherent in the laws of nature.

Nevertheless, physicists have used quantum mechanics to construct a reasonably robust theoretical framework for describing the fundamental properties of matter and the forces at work in the Universe. They hope to explain how the Universe has evolved and even how it came into existence. The 'reductionist' view is that once they formulate 'a theory of everything', it should be possible to explain the more complicated natural phenomena – how atoms and molecules behave (chemistry) and how they organize into self-replicating entities (biology). It is just a matter of time and effort. Indeed, human beings are subject to the same laws of nature as the galaxies, so eventually we may be able to make predictions about human concerns, such as the fluctuations of the stock market or the spread of an epidemic. In theory, life is supposed to be predictable.

Is this approach useful? There are considerable shortcomings. At the moment, scientists cannot even use the fundamental laws of nature to predict when the drips will fall from a leaking tap, or what the weather will be like in two weeks' time. In fact, it is difficult to predict very far ahead the motion of any object that feels the effect of more than two forces, let alone complicated systems involving interactions between many objects.

Recently, researchers in many disciplines have begun to realize that there seem to be inbuilt limits to predicting the future at all levels of complexity. It is here that chaos theory steps in to shed some light on the way the everyday world works.

Chaos theory has resulted from a synthesis of imaginative mathematics and readily accessible computer power. It presents a universe that is deterministic, obeying the fundamental physical laws, but with a predisposition for disorder, complexity and unpredictability. It reveals how many systems that are constantly changing are extremely sensitive to their initial state – position, velocity, and so on. As the system evolves in time, minute changes amplify rapidly through feedback. This means that systems starting off with only slightly differing conditions rapidly diverge in character at a later stage. Such behaviour imposes strict limitations on predicting a future state, since the prediction depends on how accurately you can measure the initial conditions. In fact, if you model such a system on a computer, by feeding in equations and numbers, just rounding off the decimal points in a different way can radically change the future behaviour of the system.

The computer revealed the subtle behaviour of chaotic systems because it can follow their trajectories over many millions of steps. This approach has exposed the abstract geometrical nature of chaos theory, in the form of computer graphics. Within the overall shape, there lies a repetitive pattern whose exquisite substructure characterizes the nature of chaos, indicating when predictability breaks down.

It is the beautiful graphics associated with chaotic systems that has made the subject so appealing to everybody who sees them. They show how quite simple equations, when fed into a computer, can produce breathtaking patterns of ever increasing complexity. Within the often lifelike forms, there are shapes that repeat themselves on smaller and smaller scales – a phenomenon called 'self-similarity'. Benoit Mandelbrot coined the word 'fractal' for such shapes. And here we have an immediate link with nature, for trees and mountains are examples of fractals.

But there is more to it than that. Scientists now have an interpretive tool for describing many of the complexities of the world, from brain rhythms to gold futures. Recently, *New Scientist* published a series of features on chaos theory and its implications. This is the complete series in book form. The articles, which have been written by some of the world's leading experts, explain what chaos is, the mathematics behind chaos, how chaos can be found in virtually every discipline from astronomy to population dynamics, and how chaos theory can be applied practically to areas such as engineering and economics. The series also looks at the role of fractal geometry in chaos theory, and how computers are now changing the way science is being done.

Computing theory is spawning ways of modelling complexity and disorder by describing information in algorithmic forms. In this way, chaos is revealing fundamental limits to human knowledge in an uncomfortable way. Because chaos implies a delicate dependence on initial conditions, a complete knowledge of a chaotic system demands knowing the position and velocity of every particle in the Universe. Even if it were possible to carry out such a task, such a measurement would disturb the system it is measuring anyway. It seems that the macroworld may have its own uncertainty principle, a result of the nature of chaotic dynamics rather than quantum probability. Chaos may even provide the Universe with an arrow of time.

Chaos also seems to be responsible for maintaining order in the natural world. Feedback mechanisms not only introduce flexibility into living systems, sustaining delicate dynamical balances, but also promote nature's propensity for self-organization. Even the beating heart relies on feedback for regularity.

Chaos also triggers our aesthetic responses. The young, in particular, have taken to the complex graphics that seem to teeter wondrously between order and randomness. Everyone enjoys that stunning mathematical object, the Mandelbrot set. It now appears on posters, T-shirts, record sleeves and pop videos. Such colourful iterations have linked mathematics with art and nature in a stimulating way. We appreciate how the spontaneous complexity generated in self-organizing systems makes a tree more beautiful than a telegraph pole. Chaos has made mathematics come alive.

I should like to thank Nigel Hawtin and Neil Hyslop for the artwork, Richard Fifield and Ian Percival for helping me to coordinate the series, and Liz Else and Peter Wrobel for help with the editing.

NINA HALL

1

Chaos: a science for the real world

IAN PERCIVAL

Traditionally, scientists have looked for the simplest view of the world around us. Now, mathematics and computer power have produced a theory that helps researchers to understand the complexities of nature. The theory of chaos touches all disciplines.

If you watch from a bridge as a leaf floats down a stream, you may see it trapped by a small whirlpool, circulate a few times, and escape, only to be trapped again further down the stream. Trying to guess what will happen to a leaf as it comes into view from under the bridge is an idle pursuit in more senses than one: the tiniest shift in the leaf's position can completely change its future course.

Small changes lead to bigger changes later. This behaviour is the signature of chaos.

Chaos is found everywhere in nature, sometimes even in the beating of the human heart. Under certain circumstances, the human heart can beat chaotically. It is controlled by natural pacemakers, which normally give it a steady, regular beat, but sometimes they do not work together properly, so that there are alternate long and short gaps between the beats. In yet more extreme conditions, the rhythm becomes irregular. A small change in the timing of one beat makes a bigger change in the next. The beating becomes chaotic, and may threaten survival. This is a good example of how regular motion makes the transition to chaos when the conditions are changed.

You can hear this transition from regular to chaotic motion in your home, by listening to a dripping tap. If you hold a fragment of kitchen foil on the sink or wash basin below the tap, while it is dripping slowly, you will hear a regular beat on the foil. Now turn on the tap very slightly, and, under the right conditions, you will hear alternate long and short beats. Give the tap another tiny turn and it will never settle down to regular stable behaviour; it has become chaotic.

Chaos is persistent instability.

Instability is part of our own environment and our culture. The situation that is balanced on a knife-edge and the straw that can break the camel's back are metaphors for life's instabilities.

Chaotic motion contrasts with the regularity that we see on a grander scale in the cosmos. People have always wondered at the order in the seasons, at the way night follows day, and at the precision with which the stars and planets appear to move across the sky. Such celestial events all have their origin in the regularity of the motion of the Earth and the other planets, explained more than 300 years ago by Isaac Newton with his laws of motion and theory of gravitation. According to these laws, the present positions and velocities of the Sun and the planets determine the positions and velocities for all past and future times.

Newton's laws of motion are the classical example of determinism, in which the future is uniquely determined by the past. When scientists look for this kind of order in the Universe, they are often rewarded. But, as we know, order is not universal; we also need to understand disorder.

One of the first mathematicians to study disorder was Pierre Simon de Laplace. He was born in Normandy and survived the French Revolution by flattering those in power. Laplace had a thoroughly Newtonian view of the Universe, yet he helped to found the theory of disorder, or probability, which describes how large numbers of events can behave in a typical way, even when the individual events are unpredictable. This happens in gambling, which was as popular in his time as it is today. Laplace even applied his thoughts on probability to the law courts. The theory of probability now helps us to estimate the spread of AIDS without understanding the detail of how it works.

So during the 19th century, there were two kinds of theory for changing systems, deterministic theories and theories of probability. The two approaches appeared to be incompatible. In the first, the future is determined from the past, with no apparent need for probability. In the second, the future depends in some random way on the past, and cannot be determined from it.

The first challenge to this picture came with the quantum theory in the 1920s and the 1930s, which is also based on calculating probabilities – theorists describe the behaviour of an electron in terms of a 'probability wave'. The second challenge came from the theory of chaos. Simple mathematical analysis shows that even in simple systems, which obey Newton's laws of motion, you cannot always predict what is going to happen next. The reason is that there is persistent instability. This often arises when an object feels the effect of more than one force.

A well-known example is a pendulum with a bob that is attracted equally to two magnets below it. When the bob moves slowly near to a point midway between the magnets, it is affected almost equally by the force from each magnet. Its future motion becomes extremely sensitive to small changes in its present position and velocity, so the motion is chaotic. Suppose the sensitivity is so great that the error in measuring its position increases by 10 times in one swing between the magnets, which is not at all exceptional. In that case, predicting its position to within a centimetre after one swing entails measuring the position to within a millimetre. To make the same prediction after four swings, its position would have to be measured to within the size of a bacterium, and after nine swings, to within less than the size of an atom. The pendulum obeys Newton's deterministic laws, but any attempt to predict its future behaviour over long times will be defeated.

This does not mean, however, that we can say nothing about the motion of this pendulum. For some initial conditions, the motion is regular, and not chaotic at all, so we can predict a long time ahead. And we can understand many properties of the chaotic motion with the help of probability theory (see 'Determinism and probability', below). But the need to use probabilities for such simple systems came as a great surprise to those who were brought up in the Newtonian tradition.

Determinism and probability: the pinball and the word

There are few better illustrations of chaos and the connection between determinism and probability than the pinball machine illustrated in Figure 1.1.

Suppose the ball is released from just above the top pin P1, and just to the left of it. Then it bounces to the left, and hits the pin P2 on the next level down. If the size of the ball and the distance between the pins are chosen well, the ball will then have an equal chance of going to the left and hitting P4, or of going to the right and hitting P5. Suppose it goes to the right. Then it has an equal chance of hitting the pins on either side on the next level, and so on, down to the lowest level. The motion of the pinball is chaotic.

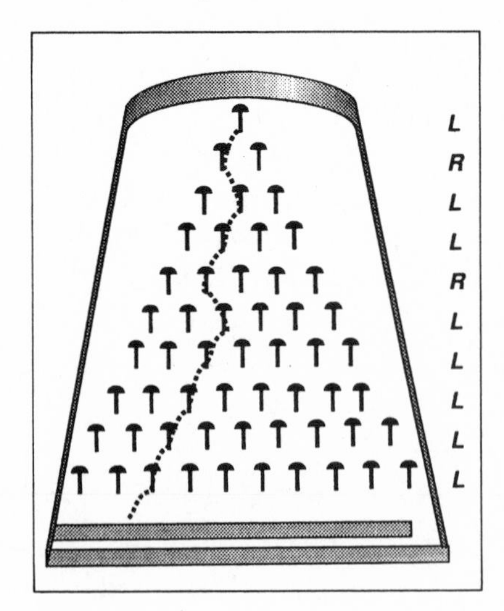

Figure 1.1

In Newtonian dynamics, the paths are usually distinguished by their 'initial conditions', which in this case means the position of the ball when it is dropped on to the first pin. But because the motion is chaotic, the path of the ball through the pins is extraordinarily sensitive to that position, so initial conditions are not very helpful.

In chaos theory, we distinguish between different paths by 'words' made up of letters. For the pinball we write an 'L' for a bounce to the left, and an 'R' for a bounce to the right.

If the numbers of layers of pins is 10, then every possible path of the ball from the top to the bottom is labelled by a 'word' of 10 letters, made up entirely of Ls and Rs, with the first letter giving the direction of the first bounce from P1, the second letter the direction of the second bounce from a pin on the second level, and so on. The word for the path shown in the Figure is LRLLRLLLLL. There are 1024 different possible paths between the pins from the top to the bottom, which is the number of such 10-letter words. In the theory of chaos, each letter and each word has a probability, just like the probability of a head or a tail in the throw of a coin, and it is this that makes the surprising connection between determinism and probability.

Distinguishing such paths or 'orbits' by words instead of initial conditions is known as symbolic dynamics. Wherever there is chaos there are words to describe it, just as for the pinball, but it is often very difficult to find them. It is worth the effort, because symbolic dynamics can tame the science of chaos.

The theory of deterministic chaos mixes determinism and probability in a totally unexpected way. Understanding the subtle unfolding of chaos in a system is helping us to describe not only the behaviour of the floating leaf, the irregular heartbeat and the dripping tap, but many aspects of our complex Universe, on both a small and a grand scale. Chaos stalks every scientific discipline.

Astronomers now use theories of chaos to model the pulsation of the early Universe and the motion of stars in galaxies, as well as that of the planets, satellites and comets of the Solar System. Nearer home, chaos helps us to study the charged particles that are trapped by the Earth's magnetism, whose escape into the atmosphere gives us the aurorae. And one of the most exciting applications of chaos theory is in studying the movements of that atmosphere that give us our weather.

Biologists also see chaos in the changing populations of insects and birds, in the spreading of epidemics, the metabolism of cells, and the propagation of impulses along our nerves. Physicists come across chaos in the motion of electrons in atoms, and of atoms in molecules and gases, and in the theory of elementary particles. Even engineers have to think about chaos because it intervenes in their designs. Chaos can frustrate those who design electrical circuits. It can lead to the loss of particles from a particle accelerator or a plasma, or to the capsize of a moored ship in a rough sea.

Some of the most beautiful examples of chaos are in mathematics, where the solutions of apparently simple problems show extraordinarily complicated behaviour. In the past, before the days of computers, this made scientists shy away from these problems, but now that we have computers to help us, the beauty of these complications is one of the subject's main sources of attraction. Chaos is a science of the computer age. And some of the elegant mathematics used to model chaos has important applications in many fields.

Nevertheless, you cannot use the theory of chaos everywhere. Science takes words and shapes their meanings to its own ends, and 'chaos' is no exception. The state of Eastern European politics may look chaotic, but you cannot study a subject of this type using chaos theory. There are many other situations that are chaotic in the ordinary sense, but not in the scientific sense of chaos.

The science of chaos is like a river that has been fed from many streams. Its sources come from every discipline – mathematics, physics, chemistry, engineering, medicine and biology, astronomy and meteorology, from those who study fluids and those who study electrical circuits, from strict and rigorous provers of theorems, and from swashbuckling computer experimenters.

Many of the new ideas in chaos were discovered independently in different fields. Researchers realized that although they were working on vastly different problems, they were employing the same kinds of mathematical techniques to deal with them. Much credit for bringing unity to the infant science belongs to those like Joseph Ford of the Georgia Institute of Technology in Atlanta, who recognized the common features early on.

The study of chaos, however, is not completely new. One early contribution came in the 19th century with a Russian mathematician called Sophia Kovalevskaya. She had received her university education in Germany, and later in Sweden became Europe's first woman professor of mathematics. Kovalevskaya made a step towards an independent theory of chaos in 1889, when she provided a mathematical definition of dynamical instability as an average of a measure of the rate of growth of small deviations. Her compatriot Aleksandr Liapunov made Kovalevskaya's definition more general, and now this average measure is universally known as the Liapunov exponent.

But there is no doubt that the most important figure from this era was Henri Poincaré, who came from Nancy in the north-east of France and worked in Paris. He has been called the last of the universal mathematicians, but he also deserves to be named as the first to study chaos. Many of the modern developments in chaos theory can be traced back to his classical work on celestial mechanics at the end of the 19th century. He knew that the methods of his day could not solve the equations governing the motion of the Solar System and that of many other similar dynamical systems. He even had a good idea why: the unstable motion of apparently simple systems can be extraordinarily complicated.

Although Poincaré was ahead of his time, and failed to convey his magnificent vision to his contemporaries, he found some remarkable connections between the practical problems of astronomy and some difficult pure mathematical problems. As a consequence, he developed a new kind of mathematics called topology, a sort of geometry that deals with continuities and connections among varying quantities. Topology has turned out to be a powerful tool for describing chaotic behaviour.

Later, other French-speaking mathematicians built on Poincaré's insight, taking up the difficult problems that his ideas had set them. For example, in the 1920s, Gaston Julia and Pierre Fatou in Paris studied a special kind of abstract chaotic motion, which led to the beautiful modern 'fractal' pictures that Benoit Mandelbrot created at IBM in Yorktown Heights in New York State in the 1970s. These stunning computer-generated geometrical forms adorn almost every popular book and article on chaos. Fractal geometry has a kind of ordered, inbuilt irregularity, and it describes the boundary between regular and chaotic motion.

Another approach that also follows from the work of Poincaré has given far reaching views of the rich and intricate landscape between order and chaos. It takes Poincaré's topological pictures of complex dynamic behaviour and uses them to probe how oscillating systems become chaotic. In 1928, Balth van der Pol, a Dutch engineer, developed a mathematical model of an oscillating electronic valve – the sort that produced whistles in a radio – based on Poincaré's dynamic topology. He also studied the beating heart. Later, in the 1950s and 1960s, the Soviet mathematician Vladimir Arnold analysed in detail the mathematics of chaos in oscillators like the beating heart. Leon Glass and his colleagues in Montreal also used similar mathematics to study fibrillations of the heart.

All these oscillating systems become chaotic because they possess an element of 'feedback'. This results from a system's reaction to opposing influences. Feedback generates complex dynamics in simple systems. P. J. Myrberg in Finland in 1958 started a line of research to plot how such dynamics evolve. This work helped Robert May, then at Princeton, to understand how populations of animals oscillate, then become chaotic, according to the supply of food. Robert Shaw and his colleagues in Santa Cruz in California used similar methods to analyse the behaviour of dripping taps. This simple system is called the logistic map. It typifies how chaotic systems behave. The most important complications of this kind of motion were unravelled by A. N. Sharkovsky in the Ukraine in 1961 and by Mitchell Feigenbaum in New Mexico by 1978. A similar system helped David Ruelle and Floris Takens in Paris to analyse the behaviour of turbulent fluids, and Edward Lorenz in Boston in 1963 to analyse the chaos in the weather.

Although the weather – the movement of air – of course obeys the deterministic laws of motion, because there are so many varying influences, the equations derived from the laws are much too complicated to solve with even the most powerful computer.

Even in apparently simple Newtonian systems where there is no dissipation of energy (so-called Hamiltonian systems), such as the motion of stars in galaxies, atoms in molecules, and molecules in gases, chaos emerges. Our group is particularly interested in studying these systems. You may be surprised to learn that there is even chaos in the Solar System.

With modifications to take account of Einstein's theory of relativity, Newton's laws are used to study not only the chaotic orbits of large bodies in the Solar System but also those of electrically charged subatomic particles in electric and magnetic fields in particle accelerators. These machines study the basic constituents of matter by making particles collide close to the speed of light. To obtain enough particles for these experiments, the particles have to be stored for as long as a day, during which time they go further than 150 times the distance of the Earth from the Sun. Over that distance, there is plenty of scope for instability and chaos, which is found in some form in all the big machines. Chaos results in particles being lost from the machines so that they are no longer available for experiments. Engineers and theoreticians use Newton's laws adapted for relativity to study the chaotic motion that they want to avoid.

In fact, the designers of accelerators recognized the problem long ago. They were among the first to use computers to calculate the chaotic orbits of particles. As early as 1953, F. K. Goward and M. G. N. Hine from the European Laboratory for Particle Physics (CERN) in Geneva used the computer at the National Physical Laboratory in London to show how the motion of a particle can change from regular to chaotic.

At that time, physicists did not really understand what was going on. A deeper comprehension came in the 1950s and 1960s, largely through the remarkable ideas of Soviet theoreticians – Andrei Kolmogorov and his brilliant school of mathematicians in Moscow, including Vladimir Arnold and Yasha Sinai, together with Boris Chirikov and his collaborators who studied theoretically the problems of designing particle accelerators at Novosibirsk.

Novosibirsk, in southern Siberia, became a major centre of

research as part of Khrushchev's plan to develop the eastern part of the Soviet Union. One of its most important scientific establishments was the Institute of Nuclear Physics, where Chirikov worked. He knew of the work of Goward and Hine, and of the problems of designing particle accelerators, but he was also aware of the potential importance of studying chaotic motion in other fields, especially in astronomy.

Many years earlier, James Jeans, the British astrophysicist, had described how stars move around in a galaxy by comparing them to gas molecules in a container. However, when astronomers used the shift of spectral lines to find the velocities of the stars, they found that the distribution of stars predicted by Jean's theory was badly wrong. Obviously, astronomers needed to approach the problem in a different way. Theories of chaos combined with the latest computers eventually provided a way forward.

In 1958, a Greek astronomer, George Contopoulos, while working in Sweden, used computers to apply Newton's laws to the motion of stars in galaxies; the results turned out to be complicated. It was difficult to see the forest for the trees.

Then in 1963, a French astronomer, Michel Hénon from the observatory at Nice, made a breakthrough. He was working at Princeton, which gave him two valuable assets: an American computer, and an able student called Carl Heiles who was looking for a research project. Interested in the same problem that Jeans had tried to solve, Hénon and Heiles set about plotting how stars behave in galaxies. At first, Hénon was surprised to get such complicated results. But later, after coming across some of the topological ideas originally hinted at by Poincaré, he suggested some simple equations for the motion of stars in galaxies, where the stars' orbits clearly showed the transition from regular to chaotic motion. These equations have been used as a paradigm for chaotic motion in Newtonian mechanics ever since.

In fact, today, chemists use these equations as a model for molecules. In 1974, Neil Pomphrey of our group employed this approach to show that chaos in classical atomic systems even has consequences at the quantum level, an active area of research. There is, however, a great deal of controversy over whether chaos really penetrates the quantum regime.

Chaos is one of the most exciting and intriguing areas of science to develop in the past 30 years. It is still very young and we still do not know how far it will change our view of the world. What is certain is that chaos theory has underlined the interdisciplinary nature of frontier research. Developments in the theory have issued from a coming together of abstract mathematics and one of the most important research tools today, the computer. This thoroughly modern marriage has, for the first time, given researchers the courage and inspiration to tackle the ever-changing complexity of the real world.

Further reading

Popular:

JAMES GLEICK, *Chaos,* Viking Penguin, 1987.

More advanced:

MICHAEL BERRY, IAN PERCIVAL and NIGEL WEISS (eds.), *Dynamical Chaos*, Princeton University Press, 1987 (paperback); Royal Society, London (hardback).

2
Chaos in the swing of a pendulum

DAVID TRITTON

Students for generations have regarded pendulums as classical examples of simple, regular motion. In fact, pendulums still hold great surprises for us.

You have only to look through this book to see that the concept of chaos has wide-ranging applications. This diversity is one of the main reasons for its importance. Yet, to understand the significance of chaos theory, it is a good idea to look at a simple system such as the chaotic behaviour of a pendulum.

Two things make the discovery of such behaviour remarkable. First, people have known about pendulums and studied them for centuries. You might think that by now we should know all about them. Secondly, pendulums are the epitome of regularity; it is very likely that the phrase 'regular as clockwork' originated when clocks were commonly regulated by pendulums. Surely, a pendulum cannot behave chaotically or unpredictably?

In fact, pendulums can show chaos in various ways. Ian Percival and Ian Stewart mention others in Chapters 1 and 4. I am going to describe the 'forced spherical pendulum'. This may consist simply of a ball on the end of a string. The ball is suspended in a way that allows it to swing in any direction. We can force the point of suspension to oscillate in a horizontal straight line, for example by driving it with a crankshaft (see Figure 2.1). The strength, or amplitude, of this driving oscillation is small compared with the length of the pendulum and the frequency is close to that at which the pendulum naturally swings.

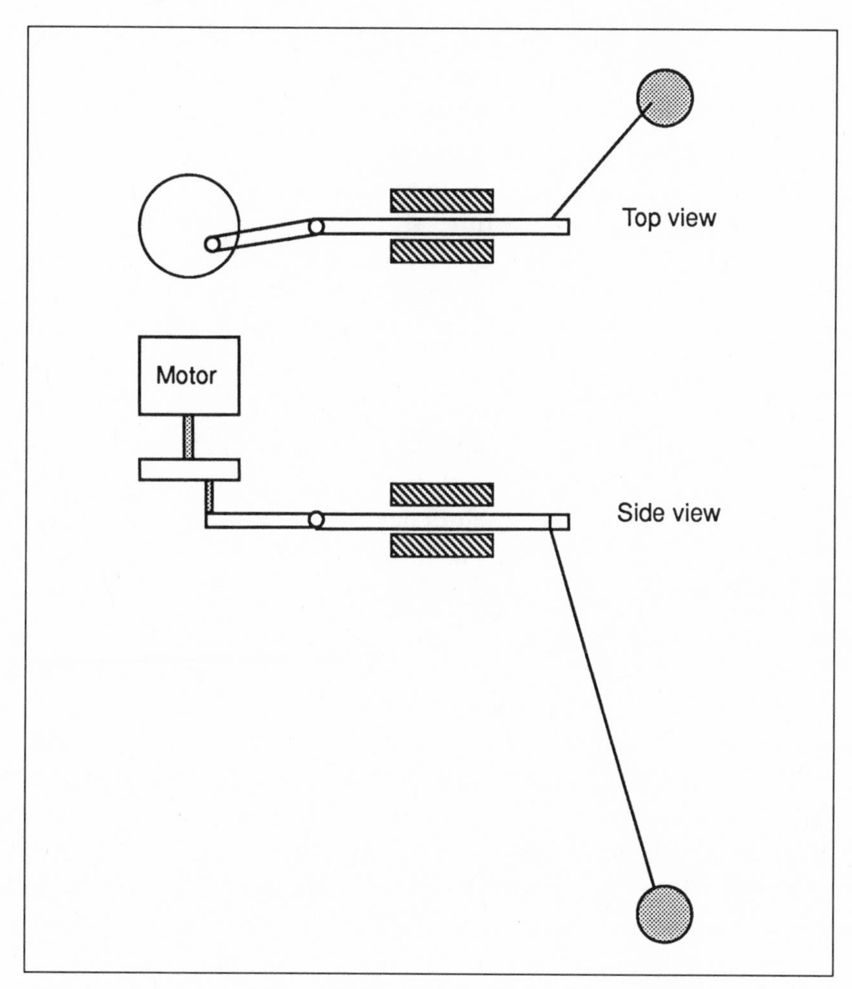

Figure 2.1 The forced spherical pendulum.

Can you guess what the pendulum will do? You will probably suppose (quite correctly) that the ball will swing parallel to the driving oscillations and that, because the frequency of the driving oscillations is close to the pendulum's natural frequency (when it swings freely), the ball may swing with a large amplitude. What is much harder to foresee – but is also true – is that the ball may also develop a motion perpendicular to the direction of the drive, and that the combined effect of both directions of motion can be very complicated. In particular, if the driving frequency is in a certain range, it is impossible to predict just how the pendulum will move; consecutive experiments under supposedly identical conditions will always produce different patterns of motion.

Yet the laws that govern the motion of pendulums are not in doubt. They are just the laws of classical mechanics, first formulated by Isaac Newton. Moreover, they are fully deterministic laws. If you know exactly how the pendulum is swinging at some instant, then in principle you can determine all its subsequent motions. The realization that such determinism in principle does not contradict the lack of predictability in practice is central to the concept of 'deterministic chaos'.

Our knowledge of the behaviour of this type of pendulum comes primarily from mathematical work by John Miles of the University of California at San Diego. My own contribution has been to construct an apparatus that demonstrates the distinction between ordered and chaotic motion.

I do not intend the apparatus to prove anything; it is Miles's theory that provides the convincing evidence that the interpretation below is the appropriate one. As a demonstration, however, the apparatus is very effective. The contrast between ordered and chaotic motion is immediately and strikingly apparent.

Incidentally, although you can make the pendulum quite easily, you need to be able to control the driving frequency very precisely. So unfortunately the demonstration does not quite come into the 'amateur scientist' category.

What you observe is the horizontal projection of the ball's motion – in other words, what you would see if viewing the ball from above. All the interesting events happen when the driving frequency is within a few per cent of the pendulum's natural frequency when it swings freely. Outside the range, the pendulum swings parallel to the drive with relatively small amplitude. Within this range, changing the driving frequency in small steps produces a variety of detailed behaviours. However, the two sharply contrasted types of behaviour each occur over a sufficiently wide range of frequencies so that it is easy to find them.

The first type of behaviour, which is not chaotic, occurs when the driving frequency is a little higher than the pendulum's natural frequency. After you switch the drive on, the ball swings parallel to the drive for a while. Because the driving and natural frequencies are close together, they 'resonate', causing the swings eventually to reach a substantial amplitude. Then the motion starts to develop a component that is perpendicular to the drive. After a complicated but repeatable sequence of fluctuations, the motion ultimately settles down to a regular pattern. The ball moves on a nearly circular path – once round the 'circle' during each period of the drive. Once this pattern becomes established, it will continue for as long as the drive keeps going.

Even this very regular motion is not completely predictable. You cannot predict whether the ball will circulate clockwise or anticlockwise. The choice between the two directions is completely random. Once the direction becomes established, however, you can predict completely the motion indefinitely far ahead.

The lack of predictability, though slight in this case, derives from the fact that the component of motion perpendicular to the drive arises through the instability of the motion parallel to it. There is a solution, with no perpendicular motion, of the equations governing the pendulum's behaviour, but this solution does not occur because any small departure from perfectly parallel motion leads to much larger departures. This is similar to trying to stand a pencil on its point. There is a solution of the equations of statics with the centre of gravity directly above the point of contact, but the slightest movement away from this position readjusts the forces so that the pencil falls. In this case, too, there is a lack of predictability: if you try to balance the pencil perfectly, you do not know whether it will fall to the left or right, forwards or backwards.

Buridan's ass died of starvation because it was exactly midway between a bale of hay and an equally attractive pail of water. It should have found a similar resolution to its problems; but who can say whether it would have eaten or drunk first?

We can understand the pendulum's motion, as I have described it so far, in terms of long-standing ideas. But when you change the driving frequency to a little below the natural frequency, there are some real surprises. What might you then see if you looked at the pendulum?

Figure 2.2 shows two of the many possible answers. Each diagram is a trajectory of the pendulum bob during several periods. Successive orbits are similar but not identical, so that after several periods there are substantial changes. Figure 2.3 demonstrates the net effect of such changes over longer times, showing an observed sequence of the bob's motion at intervals of about 15 periods, that is, around 17 seconds. Each orbit is approximately an ellipse (which sometimes becomes so narrow that it effectively forms a line), but the size, orientation and ratio of length to width all vary. Note in particular the frequent changes between clockwise and anticlockwise motion. There is no discernible pattern to these changes. Long sequences of observation, together with Miles's theoretical work, strongly suggest that there is, indeed, no pattern to be discerned. The changes occur chaotically.

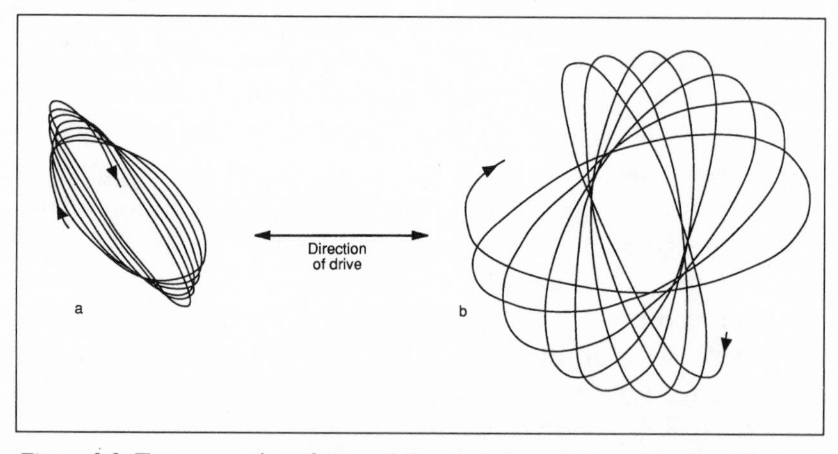

Figure 2.2 Two examples of a pendulum's bob over about seven periods of the driving motion: **a** shows an elliptical motion becoming progressively longer and thinner, with only a slight change in its orientation; in **b**, the bob's elliptical motion is progressively changing its orientation, and there is little change in the length of the bob's swing, although its path is becoming broader; **a** is part of the development between diagrams 17 and 18 in Figure 2.3, while **b** shows part of the motion that occurs between diagrams 24 and 25. (Both **a** and **b** are on the same scale.)

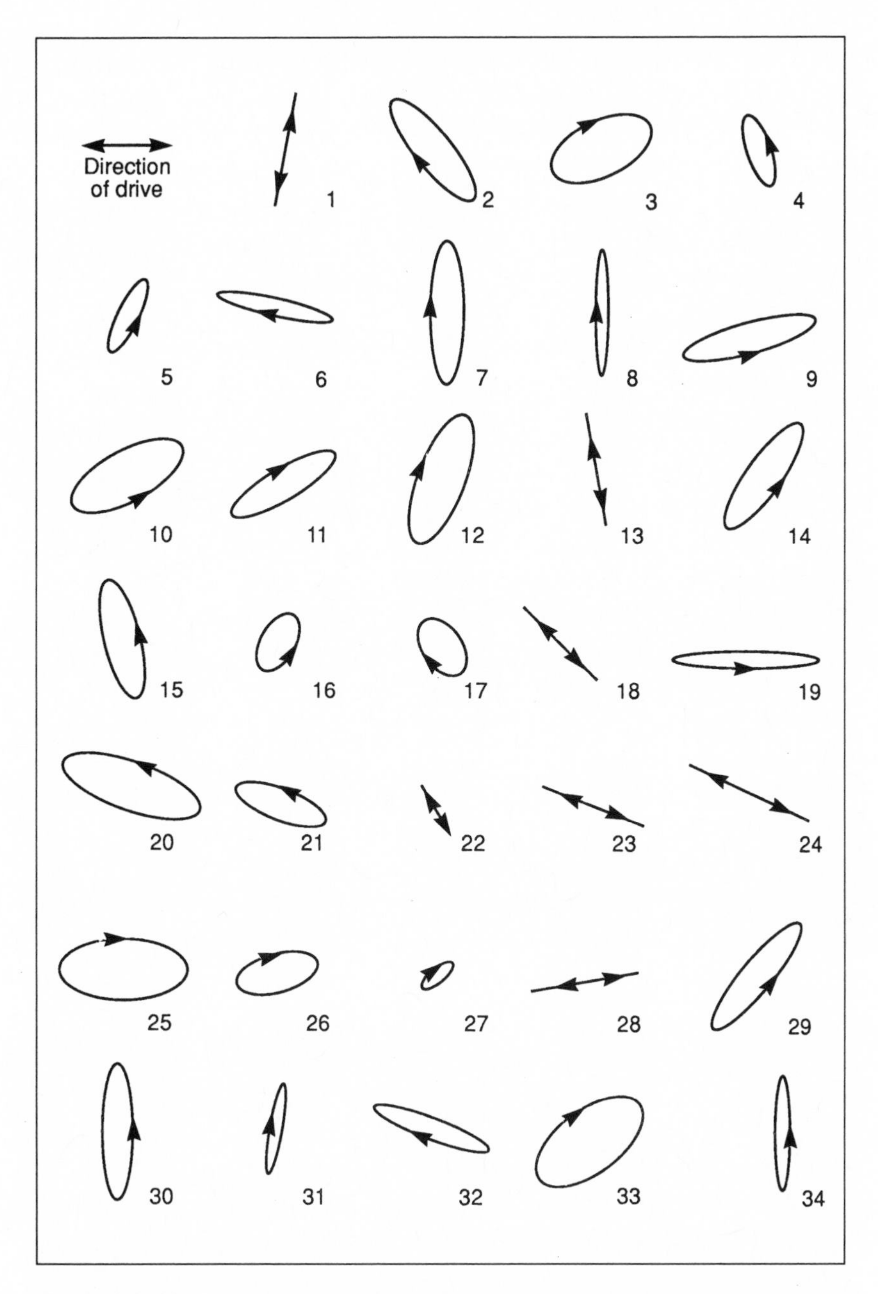

Figure 2.3 Single orbits of the pendulum's bob, observed at intervals of 15 periods of the driving motion. Note that, strictly speaking, each orbit is not closed (see Figure 2.2). A similar figure for the ordered motion described in the text would be a sequence of identical, nearly circular orbits.

What is more, another experiment at the same frequency will never yield exactly the same sequence. Shifting to a slightly different frequency – but not so different as to move into the ordered regime – may show some systematic differences. For example, more nearly circular orbits than any seen in Figure 2.3 might sometimes occur. But the pendulum will again show different detailed behaviour every time that the experiment runs.

In this case, observing the bob's motion tells us only so much about what will happen next. Suppose we obtained extremely detailed observations on the way in which the bob moved over, say, one period of the drive and fed these data into a computer, which we had programmed with the pendulum's equations of motion or with many previous observations of its behaviour. We could then attempt to predict the bob's subsequent motion. How well would our prediction agree with the actual evolution of the motion? For a while, there would be very satisfactory correspondence between prediction and observation. Sooner or later, however, the prediction would fail. The more precise are the observations that we feed into the computer, the longer the correspondence will continue, but ultimately the observed motion will always diverge from that predicted.

This 'sensitivity to initial conditions' is the key to understanding why determinism does not necessarily imply predictability. If we knew exactly how the pendulum was moving at a given time, then we could predict its future motion exactly. But we never do know anything exactly – the slightest vibration in the drive or the slightest draught in the room prevents that. The feature that distinguishes systems behaving chaotically from those that do not is that the smallest change leads ultimately to quite different detailed development.

Why do I like the spherical pendulum as a context for expounding these ideas? Two reasons – or maybe the same reason in two guises: the system is simple, and its lack of predictability is surprising. It does not overturn previous expectations to learn that there are difficulties in forecasting, say, the weather or the nation's economy. Indeed, it is the failures rather than the successes that are remembered. This is not to say that the new ideas have no impact on studies of complex systems; studies of fluid dynamics and the weather (see Chapters 5 and 6) plainly show otherwise. But, if I am asked, 'What is the biggest surprise that the chaos story has sprung?', then the answer must be that this behaviour is shown by such simple systems.

One can make this point either mathematically or physically – by looking at either equations or apparatus. Both are needed for the full picture: to know that there is a logical structure underlying what may at first appear as illogical behaviour, and to see the importance of these ideas for the real world. In the next chapter, Franco Vivaldi discusses the logistic equation; the fascination is that so much comes from so little. The spherical pendulum is actually mathematically a much more complicated system. But physically it is simple; there are no hidden mysteries in the underlying physics. Yet, just by turning the knob that controls the drive frequency, one can see the transition from ordered to chaotic motion happening before one's eyes.

Nevertheless, systems can be too simple to behave chaotically. There is a minimal requirement that the governing equations must be 'nonlinear'. The pendulum can show what this means. The period of a freely swinging pendulum does not depend on how far it is swinging. People could not have used pendulums to regulate clocks if this were not the case. This rule breaks down, however, if the swings get too big; at large amplitudes, the period does depend on the amplitude. The reason for this change is that the gravitational torque pulling the pendulum to the vertical position is proportional to the pendulum's angle to the vertical for small angles but not for large ones. This departure from proportionality is an example of nonlinearity.

Incidentally, the fact that chaotic behaviour can arise only with nonlinear systems explains why a forced spherical pendulum behaves chaotically only when the driving frequency is close to the natural frequency. Only then do the swings build up to sufficient amplitude for the nonlinearity to be significant.

Something may puzzle you. If, as is the case, such simple and such classical systems as pendulums show chaotic behaviour, why did people not observe it long ago? In all probability they

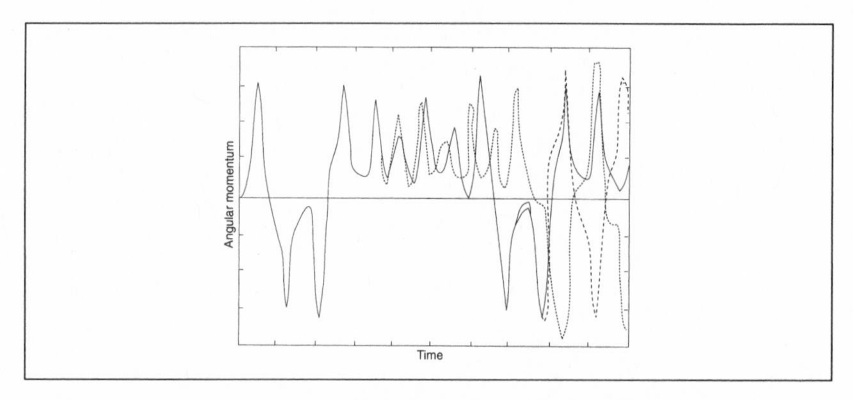

Figure 2.4 Sensitivity to initial conditions also implies sensitivity to continuous small disturbances. In numerical work, such small disturbances correspond to the finite precision (only so many decimal places) to which a computer works. The exact solution of the equations, which in principle exists, can, therefore, never be found. It can be followed for only so long; the greater the precision, the longer a numerical solution will remain close to it, but it will always ultimately diverge from it. The figure illustrates this with three plots of the variation with time of the angular momentum of the pendulum, with identical values of the drive frequency, and so on, and with the same initial conditions. The three plots were carried out with different degrees of precision: the solid line the most precise, the dashed line less precise, and the dotted line still less. We see the three cases remaining indistinguishable for a while but then diverging. Not long after the small differences appear, they quickly become very large.

did – but responded by saying 'Something has gone wrong'. They would have seen the chaos as something they should get rid of rather than interpret. (It is, for example, easy to tune the pendulum away from its chaotic regime.) Without the context of modern ideas, they could do little else.

In the first chapter, Ian Percival made the important point that the word 'chaos' used scientifically has acquired a meaning more specific than its everyday usage. It does not mean that things are just a mess. Chaotic systems usually have a rich phenomenology showing many different types of behaviour, such as 'windows' of ordered behaviour in mainly chaotic regimes and vice versa, and transitions between order and chaos by various routes. The spherical pendulum is no exception. The two types of behaviour that we have contrasted above are by no means the whole story. Only some of the 'fine structure' can be

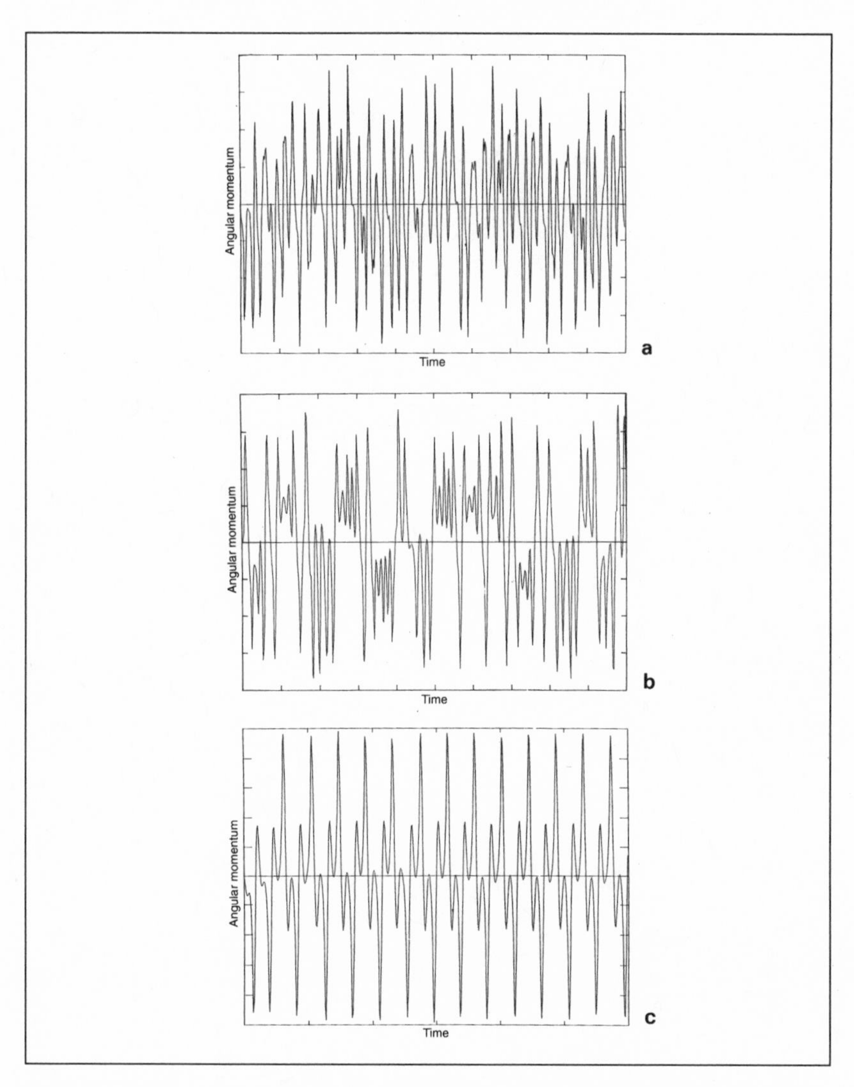

Figure 2.5 Some chaos is more chaotic than other chaos. Illustrated are three examples of the angular momentum of the pendulum against time, obtained numerically, for different drive frequencies and resistance to motion. All three are in chaotic regimes, but are very different. The chaos is apparent in **a** and **b**; **c** looks, and is, highly ordered, but every peak has a slightly different height, and these variations have all the characteristics of chaos: never repeating, sensitivity to initial conditions, and so on.

observed with my pendulum. We would need a much more refined apparatus to find the rest. Our knowledge of the details comes mainly from numerical studies of the equations of motion. Figures 2.4 and 2.5 show examples.

All this for something about which you would have supposed, a decade or two ago, there was nothing more to be learned.

So, if you meet a sceptic who says that there is little new, exciting or profound in this chaos business, suggest that he or she contemplates the chaotic motion of a pendulum.

Further reading

D. J. TRITTON, 'Ordered and chaotic motion of a forced spherical pendulum', *European Journal of Physics*, Vol. 7, 1986, p. 162.

3

An experiment with mathematics

FRANCO VIVALDI

Think of a number x, put it into a simple equation, and feed the equation to a computer. Put the answer back into the equation. Repeat the exercise and watch chaos evolve before your eyes.

If something is too large, make it smaller; if it is too small, make it larger. This is not a quotation from Mao's *Little Red Book*, but rather a simple recipe for constructing a dynamical process called feedback. It is a constant presence in our lives, and in many practical situations it is important to predict how large or small a variable quantity associated with feedback will eventually become. Common sense would suggest that it will settle down somewhere in the middle, where it is neither too large nor too small. But this answer looks suspiciously simple, and in fact it can be terribly wrong. I want to show here how feedback may turn into chaos.

The classic example comes from population dynamics, where feedback prevents populations of plants or animals from growing indefinitely. For instance, imagine that our feedback variable is the number of fishes in an ideal lake, free from pollution and fishermen. If there are few fishes, they will thrive in the favourable environment and reproduce rapidly, and the fish population will increase. But if there are too many, they will compete for food and suffer from their own pollution (an ideal lake, remember), and their number will decrease.

We can formulate this problem in an abstract setting using simple mathematics. Let us call the variable x_t, where the subscript t, a whole number, stands for the time, with the stipulation that successive times correspond to successive measurements of the variable x. Thus t does not necessarily represent the physical time, but it is rather a convenient label that orders a succession of events. In the population problem, t would label every new generation.

Assuming that $t=0$ corresponds to the present, that is, x_0 is the value measured at the beginning of the observations, then predicting the future means computing x_t. The larger the value of t, the more remote the future that we are probing. We can even let t approach infinity, a privilege denied in real forecasting.

Dynamics comes into play when we specify a rule for transforming x, which sets the picture in motion. I assume that the result of the measurement at time $t+1$ is unambiguously determined by that at time t. We can write this relation in a mathematical form:

$$x_{t+1}=f(x_t)$$

The letter f denotes a 'function' which is the way mathematicians indicate the precise relation between two quantities – x_{t+1} depends on x_t, and only x_t.

The detailed structure of the function f is of no concern at the moment, but assuming that the link between the current and the successive values of x is described by a function is no small deal. In one stroke, I have removed any ambiguity in the determination of x_t; chance, unknown external factors, noise, are not allowed to play any role here – this process is deterministic. In other words the future of deterministic systems follows from the present, without uncertainties. 'It is not like playing roulette,' it would be tempting to say. But do not say it, because we will be gambling later on, using a simple deterministic feedback system.

All information about the system is stored within the function f. Consider now a specific process, given by the following function f:

$$x_t=f_\lambda(x_t)=\lambda x_t(1-x_t), \quad \textit{Equation } 1$$

which, if you remember from school mathematics, represents a parabola. The function f depends on a certain parameter λ, which we have introduced so as to incorporate in a single description the behaviour of a whole family of feedback systems. This parameter quantifies the strength of the feedback, that is, the amount by which the feedback is correcting the value of the variable x. Geometrically, by varying λ, we vary the shape of f.

It is useful to see what f looks like for various values of λ. To do this, we can plot x_t against x_{t+1} on a plane (see Figure 3.1). All points lying above the diagonal given by $x_{t+1} = x_t$ have the property that x_{t+1} is greater than x_t, while for those below it we have x_{t+1} less than x_t. The graph of f will, therefore, lie above the diagonal for small x and below it for large x. Then it must necessarily cross that line, at least once. At the crossing point, we have $x_{t+1} = x_t$, that is x_t retains the same value in two successive measurements.

Equation 1 is a famous model, called the logistic map, originally proposed in the context of population dynamics. The

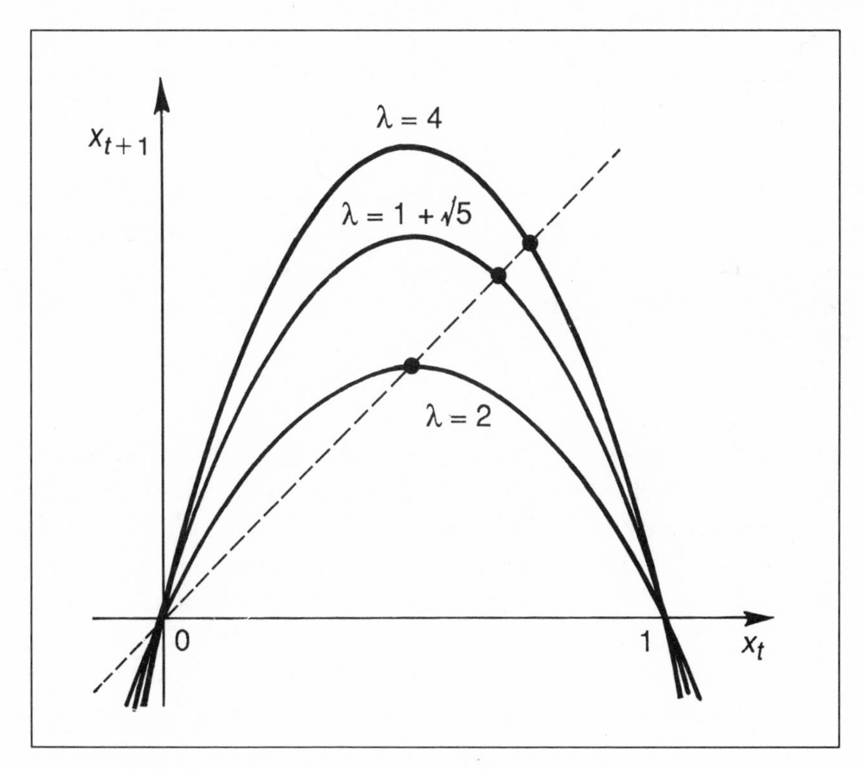

Figure 3.1 A graph of Equation 1 for different values of λ.

variable x is meant to be restricted between 0 and 1, only apparently a limitation because there is no harm in assuming that x has been suitably scaled so as to lie between these limits. Does this simple-looking problem have an exact solution? Can we derive a formula expressing x_t explicitly as a function of x_0 and λ? This would be the formula for predicting the future. The answer is yes. There is a straightforward, if naive, way of obtaining such a solution, as I shall indicate, but the result will prove to be disappointing. If you find formulas hostile and incomprehensible, bear with me for the next two paragraphs, and your worst fears will be confirmed.

I begin by letting $t=0$ in equation 1. In this way, I obtain the value of x_1 as a function of x_0 and λ, that is:

$$x_1 = \lambda x_0 (1 - x_0)$$

Now I can take equation 1 again, but with $t=1$. This gives me $x_2 = \lambda x_1(1-x_1)$, and I can substitute the equation for x_1 already obtained, to get the equation:

$$x_2 = \lambda(\lambda x_0(1-x_0)\,(1-\lambda x_0(1-x_0)))$$

One more time. From $x_3 = \lambda x_2(1-x_2)$, and using the expression for x_2 just found, I obtain x_3 as a function of x_0 and λ, that is, the future three steps ahead, as a function of the present, at the chosen value of the parameter:

$$x_3 = \lambda(\lambda(\lambda x_0(1-x_0)(1-\lambda x_0(1-x_0)))(1-\lambda(\lambda x_0(1-x_0)(1-\lambda x_0 \\ (1-x_0)))))$$

You could scarcely get excited about this accomplishment. These formulas rapidly become long and inscrutable; the expression for x_3 is already unfriendly (I could not fit it on a single line), x_{15} will fill a book and x_{30} the British Library. This is because the formula for x_t contains physically the formula for x_{t-1}, which in turn contains that for x_{t-2}, and so on. The formal solution to our problem is clearly useless for predicting anything but the very near future, and should make you ponder over the meaning of the word 'solution', an attribute these formulas do not deserve. We shall see that these equations cannot be simplified plainly because what they describe is not simple. There is chaos in the system; this problem has no solution.

What is actually happening to x_t as t increases? We ought to perform an experiment. After all, this is what people do in the physical sciences when theory fails. Rather than searching for a result valid for all x_0 and λ (the sequence of functions above), I shall choose specific values of x_0 and λ, from them compute $x_1 = f_\lambda(x_0)$, then $x_2 = f_\lambda(x_1)$, $x_3 = f_\lambda(x_2)$, and so forth, up to x_t. Manipulating numbers instead of functions makes this process much more economical. At each step, all we have to retain of the past is the previous value of x, which is one number, rather than the entire past history of all possible processes.

The outcome of this mathematical experiment will be a sequence of measurements $x_0, x_1, x_2, \ldots, x_t$, just like in real experiments. We will not have a 'formula' to pride ourselves with, but we will gain precious information by means of iterative calculations.

Computers love iteration – repeating the same task over and over again. In our case, each individual task, the computation of $f(x)$, is actually very simple, two multiplications and one subtraction in all. With each arithmetical operation taking a tiny fraction of a second, the prospect of computing $x_{1000000}$ becomes real. It can be done overnight on your personal computer, and faster than a blink on a Cray supercomputer. Moreover, in numerical experiments, the precision with which the ingredients and the results can be measured is limited only by the size and power of the computer.

To unveil the presence of chaos, I will choose a specific numerical experiment. Let $x_0 = 0{\cdot}4$, and compute x_t for $t = 1, \ldots, 15$, and for a few values of λ. Those who are familiar with a programming language will find that the bulk of the program consists of a simple iterative loop, specified by a sequence of statements such as (here in BASIC):

```
FOR I=1 TO 15
X=LAMB*X*(1-X)
PRINT X
NEXT
```

Not a very intimidating program. The results may look like Figure 3.2.

For $\lambda = 2$, the experiment brings comforting news. We witness the predicted relaxation of x, growing rapidly to a value that is neither too small, nor too large, $x = 0{\cdot}5$. This is the value at which the function $f_2(0{\cdot}5) = 0{\cdot}5$ (see Figure 3.2), which is why once x reaches 0·5, it remains there.

The outcome of the second experiment is more puzzling. The value of λ of $1 + \sqrt{5} = 3{\cdot}236\ldots$ was not a random choice. The sequence of numbers appears to approach a final regime where two distinct values of x are alternating. Had we thought about the feedback process more carefully, we could have predicted this behaviour. At this value of the parameter the feedback is strong enough to produce an overcorrection – a value of x that is too small is followed by one that is too large, and vice versa. This causes x to relax to a configuration where the opposite overcorrections balance each other precisely, and we get regular oscillations between $x = 0{\cdot}50000$ and $x = 0{\cdot}80902$.

The surprise comes from the sequence in the column on the far right ($\lambda = 4$), which you could have hardly guessed. It does not show any obvious pattern, and you might think that there was a mistake in programming, but there is no error. Successive values of x appear to be unrelated, but they are, and by just two multiplications and one subtraction; you can check. This lack of a discernible structure is not a peculiarity of the first 15 data: it will persist as long as your computer can compute. This is chaos.

It is now clear that we are in possession of a remarkable model, whose simplicity contrasts with the variety of behaviour that it can produce. So far, only the feedback strength λ has been changed, but there is more than one reason for changing the initial state as well. There was nothing special about $x_0 = 0{\cdot}4$, and we should try other values. It is perhaps even more illuminating to vary the initial state by small amounts, in order to simulate small uncertainties or errors in the measurement of the initial datum, and assess their impact on the future evolution of the system.

The second experiment is a replica of the first, but with differ-

t	$\lambda=2$	$\lambda=1+\sqrt{5}$	$\lambda=4$
0	0·40000	0·40000	0·40000
1	0·48000	0·77666	0·96000
2	0·49920	0·56133	0·15360
3	*0·50000*	0·79684	0·52003
4	*0·50000*	0·52387	0·99840
5	*0·50000*	0·80717	0·00641
6	*0·50000*	0·50368	0·02547
7	*0·50000*	0·80897	0·09928
8	*0·50000*	0·50009	0·35768
9	*0·50000*	0·80902	0·91898
10	*0·50000*	*0·50000*	0·29782
11	*0·50000*	0·80902	0·83650
12	*0·50000*	*0·50000*	0·54707
13	*0·50000*	0·80902	0·99114
14	*0·50000*	*0·50000*	0·03514
15	*0·50000*	0·80902	0·13561

Figure 3.2 The results of computing x_t in Equation 1, starting from $x_0=0{\cdot}4$, for three different values of λ.

t	$\lambda=2$	$\lambda=1+\sqrt{5}$	$\lambda=4$
0	0·**35000**	0·**35000**	0·4000**1**
1	0·4**5500**	0·7**3621**	0·9600**1**
2	0·49**595**	0·**62847**	0·153**57**
3	0·**49997**	0·7**5561**	0·5**1995**
4	*0·50000*	0·5**9758**	0·9984**1**
5	*0·50000*	0·**77820**	0·006**36**
6	*0·50000*	0·5**5856**	0·025**26**
7	*0·50000*	0·**79792**	0·09**850**
8	*0·50000*	0·5**2180**	0·35**518**
9	*0·50000*	0·80**748**	0·91**610**
10	*0·50000*	0·50**307**	0·**30743**
11	*0·50000*	0·80**899**	0·8**5167**
12	*0·50000*	0·5000**6**	0·5**0531**
13	*0·50000*	0·80902	0·99**989**
14	*0·50000*	*0·50000*	0·0**0045**
15	*0·50000*	0·80902	0·**00180**

Figure 3.3 The results of computing x_t starting from values of x_0 close to those of Figure 3.2. The numbers differing from those in Figure 3.2 are in bold.

ent values of x_0. This time we obtain the data in Figure 3.3. For the first two values of λ, the initial state was changed by 5 per cent. This would be a realistic uncertainty on the initial data, if somewhat large. The discrepancy in the early values of x_t fades away rapidly (more so in the first experiment), as the time evolution brings the system of the same final state. I invite you to take the time to try other values of x_0 in the unit interval, to convince yourself that the resulting sequences are invariably attracted to the same final regimes, a single point in the first experiment, a pair of them in the second. No wonder these sets have been given the pictorial name of 'attractors'.

For $\lambda=4$, the initial state was changed by only one part in 100 000, and we are entitled to expect a virtually identical replica of the previous experiment. But this time the error increases, and at a remarkable rate (shown in bold in Figure 3.3). By the 15th measurement, it has contaminated all available digits, making our predictive power null. This is the signature of chaos. Had we doubled the number of digits of accuracy, in other words made the accuracy 100 000 times as great, we would just have postponed the problem for twice as long. There is no way out; at some point the results of the experiments are going to become meaningless.

We can make use of this property, and transform this process into an honest mathematical toss of a coin, where the value of x smaller or greater than 0·5 will mean 'head' or 'tail', respectively. Pick the value of x_0 of your choice, and then place your bet on x_{20}.

We have arrived at the core of the issue, the realization that there are systems, even within mathematics, that are both deterministic and unpredictable. We cannot blame this failure on the influence of unknown factors, because there are none. It is rather the result of our own terminal inability to measure or represent the present with infinite precision.

People have suspected for a long time that chaos exists in dynamical systems, but it took computers to demonstrate it and assess the implications. The history of the logistic map provides a marvellous example of the interplay between theory and experiment within mathematics, something that was once a prerogative of the physical sciences.

Early theoretical results had already explained some qualitative aspects of the changing behaviour of the system as the parameter is varied (without using my awkward formulas though). They encouraged mathematicians to run extensive computations on their computers. These unveiled quantitative phenomena crucial for the understanding of chaos. A new mathematical theory originated from the computer experiments, and this process of symbiosis culminated in a computer-assisted proof of the main predictions regarding the transition between order and chaos.

Common folklore relegates mathematics to the theoretical sciences, where information and knowledge are reached by logical steps within an abstract framework, and not from experiments. In fact, mathematical discoveries are more likely to spring from the patiently acquired experience of many specific computations. The milestones of abstract thinking that have characterized the mathematics of our century have left little room for public display of the usefulness, and the charm, of mathematical experimentation.

Yet all great mathematicians of the past were eager computers, and felt little obligation to hide their calculations behind a polished façade of abstraction. Karl Friedrich Gauss, one of the greatest mathematicians, once refrained from disclosing a numerical table so as not to deprive the reader of the pleasure of computing it. The tendency towards experimentation has been particularly notable in number theory, where so many famous theorems have been inferred from numerical data, and even used before proofs were available.

Computers have added a new dimension to the experimental side of mathematics, and have made mathematical experimentation as fruitful and tangible as that of the physical world. This is particularly true in dynamics, because of the natural role of computers in iterative processes. Simple rules, like the logistic map, can disclose unexpected treasures when applied over and over again, but often the results materialize only after millions of operations.

The marriage between chaos and computers has even more profound roots. There is an intimate relationship between chaotic dynamics and the structure of the number system itself, which iteration helps bring to the surface, and which gives the science of chaos an appeal to fundamentals that few other sciences have. Extreme sensitivity of the initial conditions characterizes a chaotic system. This mathematical experiment brings the finest details of the numbers representing the initial state x_0 into centre stage. In the final analysis, the key for understanding the future is buried within the arithmetical properties of those numbers.

I cannot help wondering what mathematics would be like if the human brain could perform 10^{12} arithmetical operations per second. It would certainly be very different, and so would be our mathematical description of the physical world. Machines of that speed are now being conceived. Mathematics will change.

Period doubling and Feigenbaum numbers

The first two columns of Figures 3.2 and 3.3 in the main text are examples of what are called periodic orbits. Chaologists like to use geometrical language in describing changing values. So saying that an orbit is periodic just means that x eventually cycles back to its original value. In the case of the first column, once x settles down to a constant value, 0·5000, it returns to the same value after each successive step down the column. So the first column is said to have a period of 1. In general, we can say that if x returns exactly to its original value after T steps, the orbit has a period T and has only T distinct points. For the second column, then, the orbit settles down to period 2 as it alternates between 0·50000 and 0·80902.

For any positive value number T, there are values of λ with orbits of period T, but their arrangement is extraordinarily complicated, as you will find out if you experiment with a computer.

Among all these complications, there is one pattern that appears over and over again, and that is the pattern of 'period doubling'. Figure 3.4 below shows the periodic orbits for all values of λ between 2·5 and 3·5700. As λ

increases, the period goes from 1 to 2 to 4 and so on, through all the powers of 2. The greater the period, the faster the period doubling becomes, and the smaller the distance between neighbouring points on the orbits. For period 2048, you would need a microscope to see the structure.

The higher periods have another remarkable property, which was analysed by Mitchell Feigenbaum in the 1970s when he was at Los Alamos in the US. That is the property of 'renormalization'. When the periods are sufficiently high, the magnified fine structure for one orbit, period 2048 for example, is indistinguishable from the structure for the previous period, in this case period 1024, provided that you carry out the magnification to a precise specification: the magnification in λ should be 4·66920166 ... and the magnification in x should be 2·502908 ... Feigenbaum found that the same numbers and the same structure appear for all sufficiently smooth functions $f(x)$, whose graph has only one maximum. So these numbers, like π, are universal.

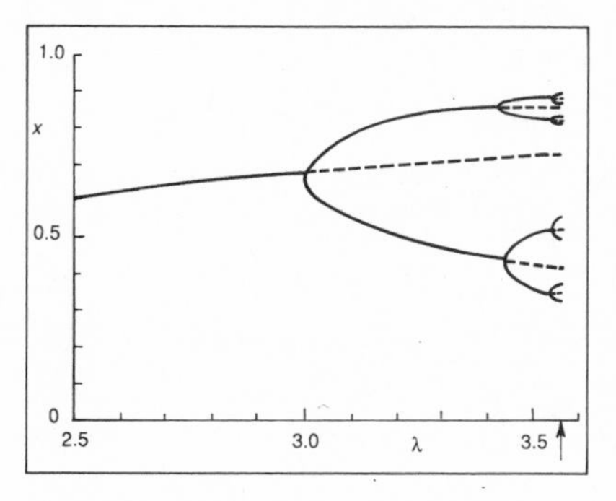

Figure 3.4 Feigenbaum doubling.

Period doubling and Feigenbaum numbers appear not only on the mathematician's computer screen but also in many kinds of natural chaos, including the dripping tap and the beating heart.

IAN PERCIVAL

Further reading

R. H. ABRAHAM and C. D. SHAW, *Dynamics – the Geometry of Behaviour*, Vols. 1–4, Aerial Press, 1982–8.

4

Portraits of chaos

IAN STEWART

The latest ideas in geometry are combining with high-tech computer graphics. The results are providing stunning new insights into chaotic motion.

A century ago, Henri Poincaré proved that the motion of three bodies under gravity can be extremely complicated. His discovery was the first evidence of what is now called chaos: the ability of simple models, without inbuilt random features, to generate highly irregular behaviour. Chaos is exciting because it opens up the possibility of simplifying complicated phenomena. Chaos is worrying because it introduces new doubts about the traditional model-building procedures of science. Chaos is fascinating because of its interplay of mathematics, science and technology. But above all chaos is beautiful. This is no accident. It is visible evidence of the beauty of mathematics, a beauty normally confined within the inner eye of the mathematician but which here spills over into the everyday world of human senses. The striking computer graphics of chaos have resonated with the global consciousness; the walls of the planet are papered with the famous Mandelbrot sets.

Why did this happen? It took three things to make chaos a household word. One was cheap computing power, rendering routine the otherwise impossible task of performing the hundreds of millions of calculations often required to take a single snapshot of a single chaotic event. The second was a growth of scientific interest in irregular phenomena, providing much of the raw fodder for the computers to gnaw on. And the third, providing the skeletal support of reason without which the nascent creature would have collapsed into a heap of quivering jelly, was a new style of mathematics, in which geometrical imagination was released from its spatial straitjacket and freed to roam through every realm of human experience.

Naturally this requires a more liberated view of what geometry actually is. There are two main styles of mathematical thought. The first is formal and remote, symbolic and abstract, centred upon the virtuoso manipulation of ever more complex formulas. The second is visual and organic, exploiting the remarkable ability of the human mind to grasp and mould shapes and forms. If the first is algebra then the second must be geometry, but geometry that is a far cry indeed from the stilting stereotype usually blamed on poor, long-suffering, misunderstood Euclid.

There is a good deal of algebra in the annals of chaos, but also an unusually large number of pictures. The role of formal calculations is that of a hard-headed critic, forcing mathematicians to check that their insights really are valid. But the insights themselves tend to be geometrical in character: you can explain them and discover them by drawing pictures. This was how Poincaré made his great discovery. From the 1960s onwards, Poincaré's geometrical approach was developed into the powerful technique in chaos theory called topological dynamics by the American School of Steven Smale in Berkeley, California, and the Soviet School of Andrei Kolomogorov, Yasha Sinai and Vladimir Arnold in Moscow. The very jargon of chaos has a strong geometrical flavour: phase space, orbits, flows, mappings, sources, sinks, saddles, attractors, bifurcations, tangles, period-doubling cascades.

Any budding chaologist who wishes not just to admire a beautiful picture, but to comprehend what it is a picture of, must come to terms with this geometrical vocabulary. To that end, let me talk you through some shapshots from the album of chaos.

The word 'space' has undergone severe metamorphosis, at least within mathematics. Conventionally, the word refers to the three-dimensional thing that we inhabit or, as my pompous dictionary puts it, 'that in which material bodies have extension'. In mathematics, 'space' can mean almost anything. A mathematician who asks: 'What space are you working in?', is not referring to the size of your office; it is a request for a description of the overall structure of the totality of objects of the kind that you are considering. From this viewpoint, a mathematician no longer studies, say, a polynomial, but an element of the space of all polynomials, This may sound like verbal hairsplitting, but it represents a considerable change in emphasis. General results about polynomials apply not just to the one you're thinking about, but to them all. And sometimes it helps to consider them all at once. In what space does chaos live?

Chaos is a dynamic phenomenon. It occurs when the state of a system changes with time. There are regular changes, the stuff of classical dynamics, and chaotic ones, and no doubt worse, which we do not yet understand. The entity that changes is some variable, or set of variables, which determines the state of the system. The values of this set of variables, at a particular instant of time, determine everything we wish to know about the system.

For example, think of a snooker ball on a table. To determine its state we need to know, at the very least, where it is (two variables, its spatial coordinates) and how fast it is going (two more). If we are interested in the spin imparted to it by the cue, we need two more variables to specify the direction of the axis of spin, and two more to measure the corresponding speeds. Even in this simplified description, a snooker ball is a pretty complicated object, requiring eight distinct variables just to specify its instantaneous state. Now let time flow. The ball moves, following the laws of dynamics. If you are a good player, it collides with a red, imparting energy that sends the red neatly into a pocket, and screws back behind the black into perfect potting position. If you are me, it jams the red in the jaws of the pocket and lines itself up on another red in perfect potting position for my opponent. Either way, the state of the cue ball changes.

That is the physical description. The geometrical view is equivalent, but with a different emphasis. Instead of thinking of the physical state of the ball, we focus on those eight quantities that determine it.

A point on a line can be specified by just one quantity, its distance to the left or right of some distinguished point, or origin. For a point on the plane, we need two such quantities, its coordinates relative to a chosen system of axes. For a point in three-dimensional space, we need three quantities. What geometry would need eight coordinates? Obviously, the geometry of an eight-dimensional space. Such a space is 'fictitious', in the sense that it exists in the mind rather than as the real object; but it faithfully represents a genuine physical entity: the set of all possible states of the snooker ball. We call it the phase space of the ball (The word 'phase' is a hangover from times past and, like most hangovers, should be taken with a dose of salt).

To define an eight-dimensional space, of course, the mathematician turns the entire idea on its head: such a creature is simply a set of lists of eight numbers. But by using the word 'space' it becomes possible to exploit analogies with ordinary space, for example, the idea of continuous motion. If a point in ordinary space moves, it traces out a curve. When a snooker ball follows the laws of dynamics, its eight coordinates move continuously, and so give rise to a curve in eight-dimensional phase space. This is not the same as the actual physical path followed by the ball on the two-dimensional space formed by the snooker table, because this takes account only of two of the variables, the space coordinates. The path in phase space is, so to speak, what happens to the physical path when we also keep track of the other six variables.

You can do the same kind of thing for any system that evolves in time. Each state is determined by a set of numbers. Thinking of these as coordinates, we obtain a multidimensional space whose points correspond to all possible states, which we again call the phase space of the system. Eight dimensions are hard to visualize, which, incidentally, is one of the things that makes snooker hard. But practice and experience help. Inside the good player's head in some form there is a level of understanding of snooker's eight dimensions which greatly exceeds that to be found in mine. One way to visualize more than three dimensions is to plot them into groups of two or three. If you look vertically downwards from an aeroplane, the scenery below is compressed into just its two horizontal coordinates. So forgetting all but a few dimensions is like looking from some specific direction, projecting the space on to a lower-dimensional screen.

Can you apply phase spaces to the ups and downs of real life? Fish populations often fluctuate in roughly periodic cycles, repeating pretty much the same changes over and over again. Why? When the First World War ended, Vito Volterra, an Italian mathematician who had spent the war developing dirigibles, tried to find out. His answer was a system of equations: here is a verbal version of what they say.

Imagine a sea in which a small number of predators, sharks say, feed from a large number of prey – shrimps. I employ these terms for reasons of alliteration rather than biological accuracy. The shark population is limited solely (shrimply?) by the number of shrimps. Suppose we start with a lot of shrimp and a few sharks. The sharks guzzle shrimp like science writers guzzle sandwiches at a press reception, and the population explodes exponentially, until the number of shrimp begins to decline sharply. No longer can the food supply support such a vast number of sharks, so the sharks die off. The shrimp population, now safe from predation, begins to rise again ... and we are back where we started.

Volterra's equations made this process precise and analysable. To interpret his results geometrically, we work in shrimp-shark space, otherwise the humble plane, which is the phase space for his model (see Figure 4.1). Every point in shrimp-shark space represents two quantities: the number of shrimp (its horizontal coordinate) and the number of sharks (its vertical coordinate). Thus the point with coordinates (243, 175 643) represents a state in which there are 243 sharks and 175 643 shrimp. Incidentally, Volterra's model assumes that the populations need not be whole numbers, because what it really works with is a fraction of some hypothetical maximum population. As time varies, the point representing the two population values moves round and round in a roughly circular trajectory. In Volterra's model, the populations always return to the identical pair of numbers, and the trajectory closes up into a loop: this is the geometer's way of spotting periodic cycles.

We made the step from a single state (point) to a trajectory (path) by considering the sequence of states into which an initial state develops. But we can take an even more 'global' viewpoint, by looking at all possible initial states. Each of these generates its own trajectory, so we get not just one curve, but a whole family of curves. For Volterra's equations, these curves are all closed loops, nested inside each other like the rings in the trunk of a tree. This system of all possible paths along which the system can evolve is called its 'phase portrait'. In practice, we draw only a small number of the paths, leaving it to the eye to interpolate the rest.

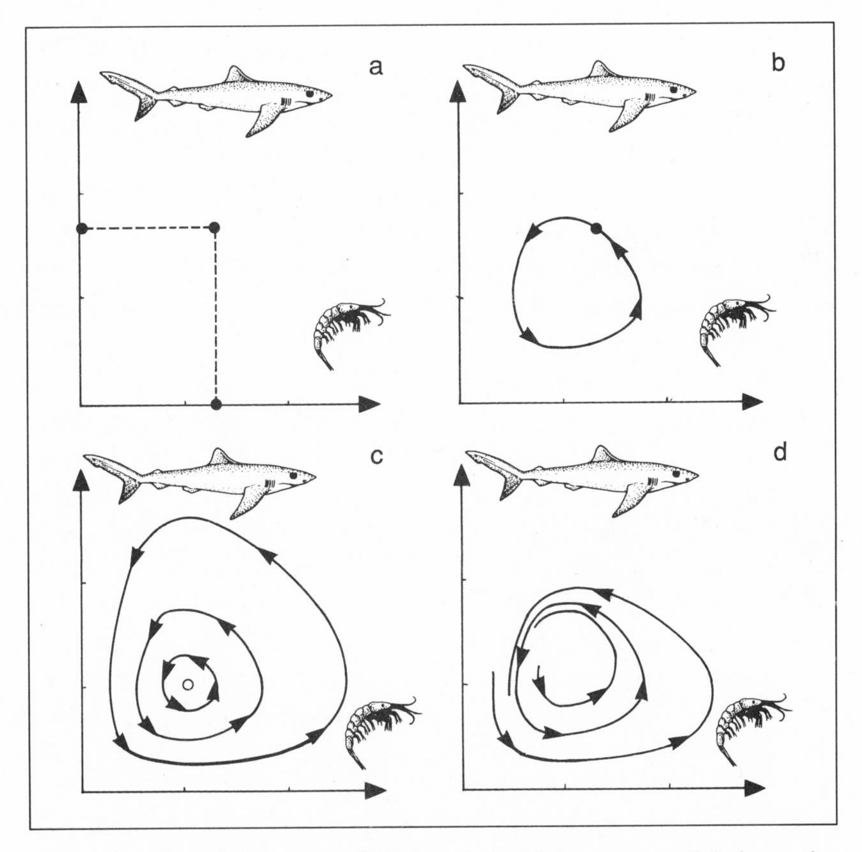

Figure 4.1 The phase space of Volterra's predator-prey model: in **a**, the coordinates of a point represent the initial values of two populations, shark and shrimp; in **b**, the trajectory representing development from an initial pair of population values is a closed loop; in **c**, the phase portrait, representing all possible developments, consists of nested loops; in **d**, a modified model has an attractor towards which all other trajectories spiral. The attractor characterizes the long-term behaviour of the system. Such phase portraits, extended to many dimensions, are helping researchers to understand complicated dynamics in evolving systems.

Poincaré realized that there are many qualitative restrictions on such a family of curves. For example, if the phase space is a plane, then every closed loop must contain a fixed point. That is, the presence of periodic solutions implies the presence of steady states too. He found more powerful arguments of the same type, in which the topology of the phase portrait had dynamical implications. It took about 80 years to turn those early insights into the sophisticated and effective techniques of today's topological dynamics.

In practice, Volterra's model is not terribly good. In particular, it predicts that any initial pair of populations will produce a cycle, eventually getting back to the same values. That is, there are many possible cycles. In practice, the shark-shrimp population tends to settle down towards some definite cycle, no matter what the initial values are. All starting values get 'trapped' into this one endless cycle. Pictorially, all paths other than the cycle (and the fixed point inside it) spiral inwards or outwards. Quantitatively, all population graphs fluctuate up and down, but settle towards the same basic waveform.

Thus, from the rather complex system of curves that forms the entire phase portrait, we can select just one curve, which corresponds to the long-term evolution of the system. No matter where you start from, if you wait long enough, the system will follow this single trajectory to as high a degree of approximation as you wish. Any region of phase space with this property is known as an attractor. Usually, a system does not explore the whole of phase space: instead, the dynamical laws pick out small regions, and almost all of the long-term motion takes place solely within those regions. Any point that starts outside those regions is 'attracted' into them.

What do attractors look like? Hundreds of thousands of calculations have been carried out in classical mathematics, determining the changes in time corresponding to various systems of equations. The great majority of the answers fall into two classes: steady-state behaviour (nothing happens) and periodic (the same thing repeats over and over again forever). It is chastening to realize that from the qualitative viewpoint these myriad calculations reveal the presence of just two distinct forms of attractor. One is a single point, corresponding to a steady state; the other is a closed loop, corresponding to periodic motion.

These are rather simple ingredients from which to cook up our mad Universe. Is there anything else?

A more complicated type of motion that also occurs, though less often, in the traditional literature, is quasiperiodic motion. Here two (or more) distinct periodic motions, of unrelated periods, are combined. The result is a motion that almost repeats, but never quite gets back to the exact starting state. Imagine an astronaut in a space capsule orbiting the Earth, testing out the theory that there isn't room to swing a cat in a space capsule. The cat goes periodically around the astronaut, the astronaut in her capsule goes periodically round the Earth, the Earth goes periodically round the Sun, the Sun goes round the Galaxy, and the Galaxy goes periodically round the Restaurant at the End of the Universe. That's five distinct periodic motions, all combined. The combined motion repeats exactly if and only if the motions are resonant: there is some period of time that is an exact whole number multiple of each of the separate periods. Usually this doesn't happen.

The geometrical picture for quasiperiodic motion in phase space is a curve that combines two or more different 'circular' motions. The combination of two circles yields a torus, a dough-nut-shaped object, complete with central hole (see Figure 4.2). A quasiperiodic orbit winds itself round a torus like thread on a

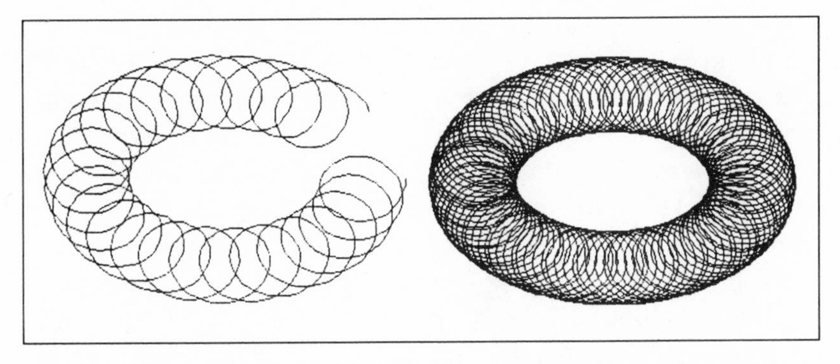

Figure 4.2 The combination of two independent periodic motions produces a torus. A quasiperiodic orbit winds itself around the torus like thread on a spool.

spool. For a combination of more than two periodic motions, we use multidimensional tori (see 'The five-dimensional pendulum picture show', below). Indeed, we can even think of a single point as a zero-dimensional torus (combine no periodic motions) and a closed loop as a one-dimensional torus (combine one periodic motion), so the total activity of classical dynamics boils down to . . . nothing but tori. Hundreds of thousands of calculations, and every one a torus.

It's all pretty cosy, really; neatly wrapped up into the GUT, the Grand Unified Toroid. All the world's a torus, and all the men and women merely quasiperiodic trajectories . . . or maybe not. Did classical mathematics miss something? Indeed it did. Classical mathematics had rather limited tools. It wanted solutions that could be specified by a tidy formula, so it therefore concentrated on equations that could easily be solved by a tidy formula. Unfortunately, most cannot, and within this silent majority there lurk innumerable attractors of a distinctly less cosy form than your friendly neighbourhood torus. They are called strange attractors. That does not mean they are in any way unusual; indeed the only thing unusual about them is that they are unusually common. It means that nobody understands them very well.

One of the first was found by meteorologist Edward Lorenz, at the Massachusetts Institute of Technology in 1963. The Lorenz attractors look rather like a mask with two eyeholes, but twisted so that the left- and right-hand sides bend in different directions (see Plate 3). How can it lead to chaos? The answer is geometrical, and simple. Trajectories wind round the two eyeholes of the mask, where both eyeholes merge together. Whichever direction you have come from, you still have a choice. Moreover, points that start close together get stretched apart as they circulate round the attractor, so they 'lose contact', and can follow independent trajectories. This makes the sequence of lefts and rights unpredictable in the long term. This combination of factors, stretching points apart and 're-injecting' them back into small regions, is typical of all strange attractors.

Another typical feature is that they are fractals, that is, they have complete structure on any scale of magnification. It may

The five-dimensional pendulum picture show

Recently, Anne Skeldon and Tom Mullin of the Nonlinear Systems Group at the Clarendon Laboratory, Oxford, collaborated with David Pottinger at the IBM UK Scientific Centre in Winchester to investigate the best way of using three-dimensional computer graphics to make phase portraits.

The project involved studying the motions of a simple system of coupled pendulums. One pendulum hangs down, and is allowed to swing from side to side. The second pendulum is attached to the bottom of the first pendulum but swings in a direction at right angles to the first. The pivot point of the top pendulum is driven up and down in a regular way by a motor at a certain frequency (see Figure 4.3). The researchers were interested in studying the patterns produced as the two pendulums moved around, when the frequency of the motor was changed.

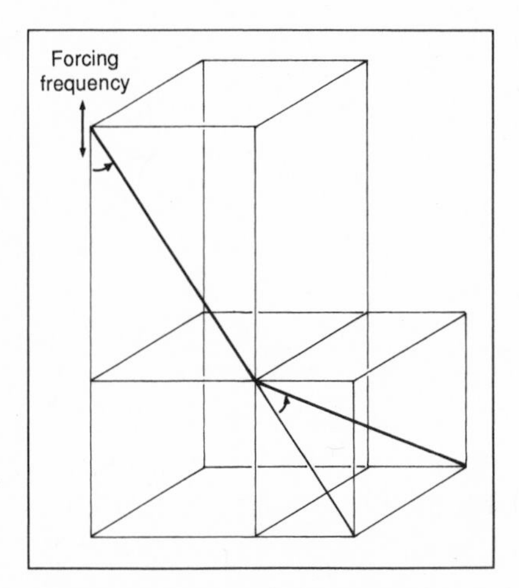

Figure 4.3 Coupled pendulums driven by a motor.

Like the astronautical system mentioned on p. 51, the system is a complex assembly of periodic motions. These are described in terms of five coordinates. They are the velocity and angle of the top pendulum, the velocity of the drive and the velocity and angle of the bottom pendulum.

To obtain the phase portraits, the researchers took the differential equations describing the periodic motions of the pendulums, then put in starting values for all the coordinates and calculated the values for a later time on a computer. The computer repeats the process through a series of time-steps until it has an adequate numerical description of the system's dynamic behaviour. In other words, there is access to the complete trajectories of all five coordinates.

The computer now has enough information to provide a phase

portrait. Because there are five coordinates, the phase portrait must be in five-dimensional phase space. The next trick was to find the best way of displaying five dimensions on a two-dimensional computer screen. A method was chosen that highlighted the important qualitative features.

Plates 4 to 9 show the projections obtained by picking three of the five coordinates and plotting the resulting trajectories. What we get are two attractors for the system. Each attractor is a torus, coloured differently to indicate that it is distinct. Both attractors are for the same driving frequency of the motor. The one that is obtained depends on the starting values of the coordinates.

The apparent intersection of the two tori is an artefact of the projection, because, remember, the system is in five dimensions. If we change the driving frequency slightly, the tori become distorted and move closer together as shown in Plate 5. On reducing the frequency further, the two separate tori join through a 'gluing' bifurcation to become one (monochrome) attractor as in Plate 6.

To understand the five-dimensional attractor better, colours are assigned to the trajectories to indicate a fourth dimension. In Plate 7, the projection of the attractor is chosen in the same way as above and is coloured according to the sign of the fourth coordinate. This tells us on which part of the cycle the pendulum changes direction.

Plate 8 is a similar attractor, but is coloured according to the magnitude of the fourth coordinate. The green band indicates where one of the pendulums is moving more slowly and successfully shows that the 'gluing' process happens in this part of the motion. Plate 9 shows the two

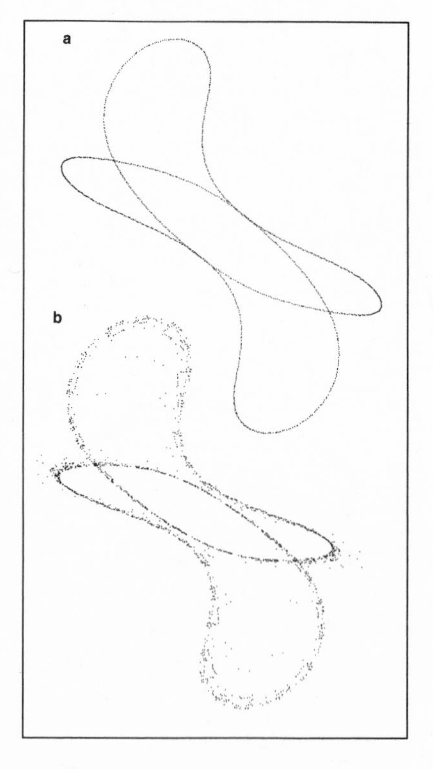

Figure 4.4**a** and **b** Poincaré sections of the joined torus showing how a slight change in the driving frequency causes the motion to become chaotic.

separate tori, coloured in the same way as in Plate 8, just before the 'gluing' process has taken place.

Finally, in Figure 4.4a, a Poincaré section of the joined torus is shown; notice that the surface is smooth. If, however, we change the driving frequency by a tiny amount, the motion becomes chaotic, as shown in Figure 4.4b, and the torus begins to break up. Thus there is no longer a smooth surface associated with the phase portrait and representing it will require some interesting new graphical techniques.

---

appear that the Lorenz attractor is a smooth surface; if you work closely enough, you'll find that it has infinitely many layers, like an extreme version of puff pastry.

Now dealing with this kind of geometry in high dimensions is hard. Almost any trick is worth having if it reduces the number of dimensions, even if it loses information. One of the most useful tricks in the topological book is the formation of a Poincaré section, which lowers the dimension by one: from three dimensions to two, or from four to three.

The so-called Rössler attractor, for example, resembles a Möbius band and lives in three-dimensional space. Trajectories loop round and round the band. Because of the way the band folds up, the precise position across the width of the band varies

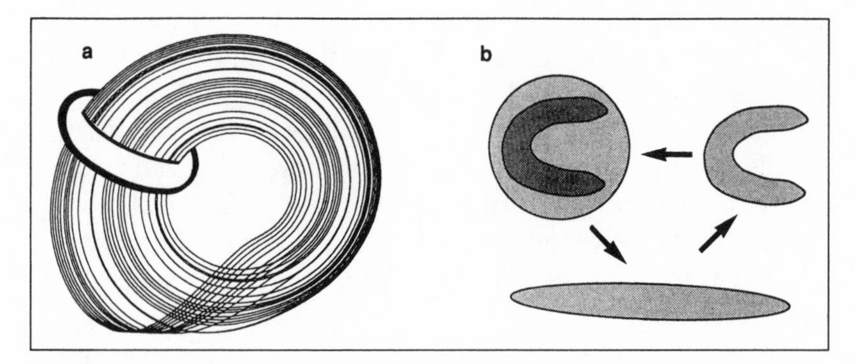

Figure 4.5 The Rössler attractor: **a** is a Poincaré section, and **b** is the corresponding Poincaré mapping. The dynamics takes the circular section, flattens it, bends it into a U-shape, and replaces it within its original outlines. Iteration of this process produces an infinitely layered attractor, which when swept round in a loop reconstructs the original Rössler attractor.

chaotically. Thus the direction across the band contains the main part of the chaos; that round the band is much tamer.

Imagine a paper hoop stretched out across the band (see Figure 4.5). Any given trajectory jumps through the hoop, meeting the paper in a single point; then wanders round the attractor, then jumps through the hoop again at some other point. This process defines a mapping from the paper to itself, that is, a rule assigning to each point of the paper another point, its image. Here, the image of a given initial point is just its point of first return.

The paper hoop is a Poincaré section, and the 'first return' rule is its Poincaré mapping and can be described as follows. Stretch the original sheet of paper out to make it long and thin; bend it into a U-shape, and replace it within its original outlines. We obtain a kind of stroboscopic view or cross-section of the dynamics of the full system by iterating or repeatedly applying the Poincaré mapping. We lose some information – such as precisely what happens in between returns to the hoop – but we capture a great deal of the dynamics, including the distinction between order and chaos.

It is very easy to iterate a mapping, even a complicated one, on a computer. So Poincaré mappings can be used to simplify quantitative calculations too, or to present experimental results in a more meaningful way. Most equations describing physical systems contain a number of adjustable values, or parameters. For example, think of a ping-pong ball sitting in a bowl on a record-player turntable. The speed of the motor is an adjustable parameter, affecting the precise behaviour of the ball. At low speeds, the ball sits at the bottom of the bowl in a steady state – that is, a point attractor. But at some critical speed, it climbs up the side of the bowl and goes round and round periodically in a circle. The topology of the attractor changes from a single point to a closed loop.

Any change in the qualitative nature of the attractor is called a bifurcation. More complicated bifurcations can create strange attractors from conventional ones (see Figure 4.6). Thus bifurcations provide a route from order to chaos, and it is by studying such routes that most of our understanding of chaos has been obtained. For example, if a fluid is pumped along at faster and faster speeds, it makes a sudden transition from smooth flow to turbulent flow. At least in some specific cases this transition is accurately modelled by bifurcation from a torus to a strange attractor. Turbulence is topological.

Poincaré's great vision was that dynamics could be made visual. It has taken the efforts of three generations of math-

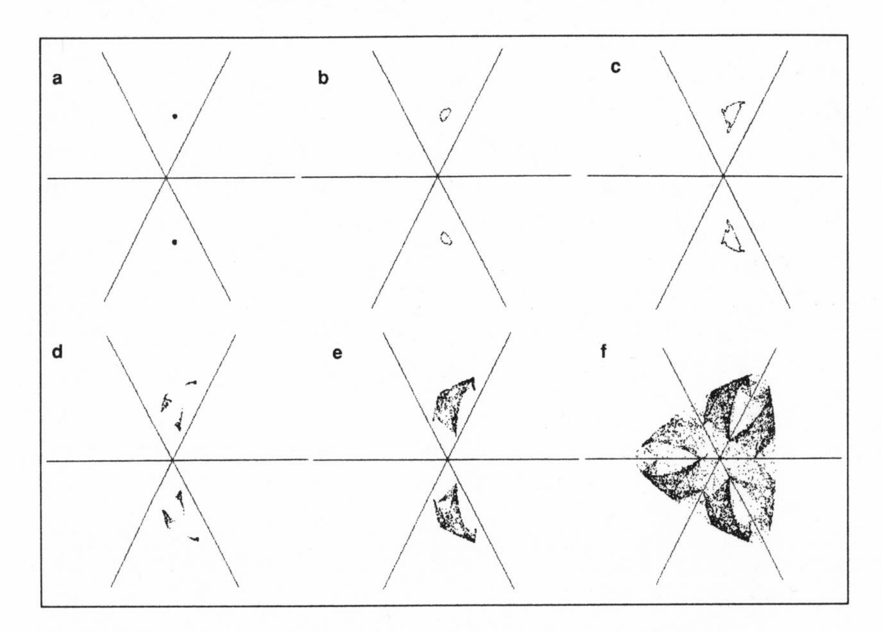

Figure 4.6 A series of bifurcations in the attractor of the mapping of the equation:

$$[x, y] \rightarrow [(2x^2 + 2y^2 - p)x - \tfrac{1}{2}(x^2 - y^2), (2x^2 + 2y^2 - p)y + xy]$$

which involves an adjustable parameter p. (This is a much more complicated version of the one-dimensional map described by Franco Vivaldi in the last chapter.) The mapping has the symmetry of an equilateral triangle, and straight lines show the axes of symmetry. Simple electronic circuits formed by coupling three identical oscillators behave in this kind of way, and are currently under study in the University of Warwick's Nonlinear Systems Laboratory. In these circuits, p represents the strength of the coupling, and the behaviour becomes increasingly chaotic as p increases. In **a**, the parameter $p = 1{\cdot}5$; the attractor is a point pair, where the state flips between two possible values. In **b**, $p = 1{\cdot}8$; the attractor bifurcates to form two closed loops, representing quasiperiodic behaviour. In **c**, $p = 1{\cdot}87$; the loops start to collapse. In **d**, $p = 1{\cdot}9$; a strange attractor with six components starts to form. In **e**, $p = 1{\cdot}92$; the components merge into two. In **f**, where $p = 1{\cdot}93$, there is a symmetry-increasing explosion, or 'crisis', caused by several attractors colliding.

ematicians to realize this vision. I have shown you just a few snapshots, the most basic, from the album of chaos. They include a single trajectory, a phase portrait, an attractor, a bifurcation. The information conveyed is different in each case: how a given initial condition develops, how all possible initial conditions develop, what the long-term behaviour is, or how that behaviour changes as parameters are varied. The great discovery of chaotic dynamics is that apparently patternless behaviour may become simple and comprehensible if you look at the right picture. Thus visual imagination, one of the most powerful attributes of the human mind, is brought to bear. Not only are the portraits of chaos strikingly beautiful: they encapsulate an enormous quantity of information in a single coherent structure. In the world of chaos, a picture is worth a million numbers.

Further reading

IAN STEWART, *Does God Play Dice?*, Basil Blackwell, 1989. 'Chaos', *Scientific American*, December 1986, p. 38.

JAMES P. CRUTCHFIELD, V. DOYNE FARMER, NORMAN H. PACKARD and ROBERT S. SHAW, 'Chaos' , *Scientific American,* December 1986, p. 38

ROBERT L. DEVANEY, *Chaotic Dynamical Systems,* Addison-Wesley, 1989.

5

Turbulent times for fluids

TOM MULLIN

Babbling brooks and bracing breezes may please poets, but they bother physicists. These natural examples of turbulence are difficult to analyse mathematically. Now, theories of chaos combined with some simple laboratory experiments may provide some answers.

Turbulence is probably the most important and yet least understood problem in classical physics. The majority of fluid flows that are interesting from a practical point of view – from the movement of air in the atmosphere to the flow of water in central heating systems – behave in a disordered way. Turbulence has always worried physicists because it is so difficult to model. In 1932, the British physicist Horace Lamb told a meeting of the British Association for the Advancement of Science: 'I am an old man now, and when I die and go to Heaven there are two matters on which I hope for enlightenment. One is quantum electrodynamics, and the other is the turbulent motion of fluids. And about the former I am really rather optimistic.'

Nearly 60 years later, the fundamental nature of turbulence remains a mystery, although new and exciting developments continue to emerge. One of these developments is the application of some of the ideas of chaos. Fluid dynamics has proved to be a useful test bed for mathematical theories modelling the transition from order to chaos.

For more than a century, fluid dynamicists and mathematicians have been trying to understand turbulence in fluids, by analysing the mechanisms that generate disordered motion. As a starting point for their investigations, they have used a set of equations that describe how both liquids and gases move. These equations, which were developed independently by Claude Navier and George Stokes in the first half of the last century, are based on Newton's laws of motion, so they are deterministic. You might, therefore, expect that by putting a set of measurements – velocity, time, and so on – into these equations you would be able to produce solutions that describe the motion of the fluid for all time.

Unfortunately, this is not so. Reading the earlier chapter by Franco Vivaldi on the logistic map, you may have been alarmed to discover that even extremely simple sets of deterministic equations, when allowed to evolve over time, quickly give chaotic answers. The reason is that the equations are 'nonlinear' (the variables in the equations are not directly proportional to one another but vary as the square or some higher power). The Navier-Stokes equations are also nonlinear and much more complicated, so you should not be surprised to find that chaos, or turbulence, is the rule rather than the exception.

Furthermore, even with today's most powerful computers, we cannot usually solve these equations. Take a simple example, such as the flow from a tap. The parameter that describes the flow is called the Reynolds number, R, after the British engineer Osborne Reynolds. It is the ratio of the inertia of the fluid, as defined by its mass and velocity, to the stickiness, or viscosity, of the fluid. This means that R is small when the viscosity is high and the velocity low. A high viscosity tends to damp out any disturbances in the motion of the fluid, which are dissipated as heat. You do not usually see turbulence in molasses.

So, for small values of R, when the tap is only slightly open, the water falls smoothly as shown in Figure 5.1a below. We describe this as laminar flow, because the water flows in parallel sheets. Even calculating this simple state of affairs would test the limits of a supercomputer if we started from the full Navier-Stokes equations. What is more, laminar flow is not usually found in nature, so it does not have much practical value.

If we increase the Reynolds number by opening the tap further, then we obtain the flow shown in Figure 5.1b. This disordered

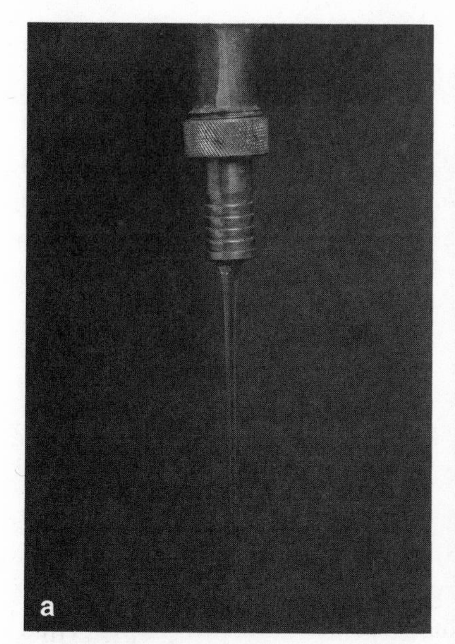

Figure 5.1 When the tap is only slightly open, as in **a**, water falls in a streamlined way. Even so, calculating what is happening is extremely difficult. Turning the tap further causes the water to flow turbulently, as in **b**. It is virtually impossible to solve the mathematical equations that describe this state of affairs.

motion is called turbulent and is the most typical fluid flow encountered. In this case, it is impossible to calculate the flow from the Navier-Stokes equations. Engineers can, however, make some progress with *ad hoc* models that capture some of the essential features of the flow, but only rarely do these models apply to more than one situation. One way to understand how and why turbulence happens is to study the fundamental processes causing it. The transition to turbulence in water from a tap is as sudden as it is in the case for flow in a pipe – which was the problem on which Reynolds carried out his pioneering work. Both these situations are extremely difficult to control in the laboratory, so, although they are important practically, exacting laboratory set-ups with taps make unattractive experiments.

Instead, some physicists prefer to study a classical system that is much easier to work with. The Taylor-Couette system, first studied at Cambridge University by Geoffrey Ingram Taylor in the 1920s, shows the onset of disordered motion through a sequence of stages. In this experiment, we look at the movement of fluid in a gap between two concentric cylinders where the inner cylinder rotates and the outer one is kept still. At small values of the Reynolds number *R*, the motion is mainly in concentric circles around the axis of the cylinders, except for some inevitable motion in other directions near the ends of the apparatus. If we increase *R* by speeding up the rotation of the cylinder so that the fluid moves faster, a rather strange thing happens. At a certain critical point, a secondary motion appears superimposed on the concentric motion. This secondary motion is in the form of cells that look like stacked jelly rolls (see Figure 5.2). We have made the flow visible by adding metal particles that reflect the incident light. The photograph, taken with a time exposure of four seconds, shows the paths of the spiralling particles caused by the secondary motion superposed on the main flow round the cylinder.

If we now view a cross section through the flow, then we can see the circular motion shown in Figure 5.3. This is done by illuminating a section through the cylindrical gap with a slit of light from the side, and viewing the flow in a direction at right angles to the light. Compare this experimental result with the picture above, which shows the stream lines calculated from the Navier-Stokes equations by Andrew Cliffe using the Cray supercomputer at Harwell Laboratory.

When we increase *R* a little more, then waves appear in the cells as shown in Figure 5.4. The waves travel around the cylindrical gap at some fraction of the speed of the inner cylinder (typically in the range between 0·1 and 0·5). If we were to measure the velocity at a point in such a flow using a laser as a probe, then the change in velocity over time would appear as a simple wave (a sine wave) when shown on an oscilloscope.

If we now increase the Reynolds number above a second critical value, then another frequency will appear, which is not related to the first. In 1944, the famous Soviet physicist Lev Landau had suggested that further increases in *R* would produce more and more frequencies in the flow with each new wave

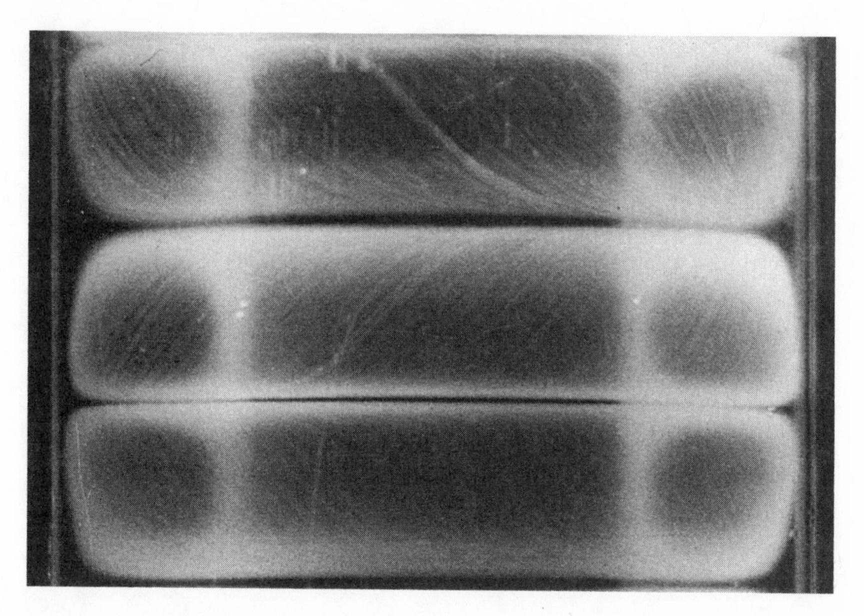

Figure 5.2 A front view (at 90° to Figure 5.3) of circular cells that appear in a Taylor-Couette system as fluid flows between two cylinders.

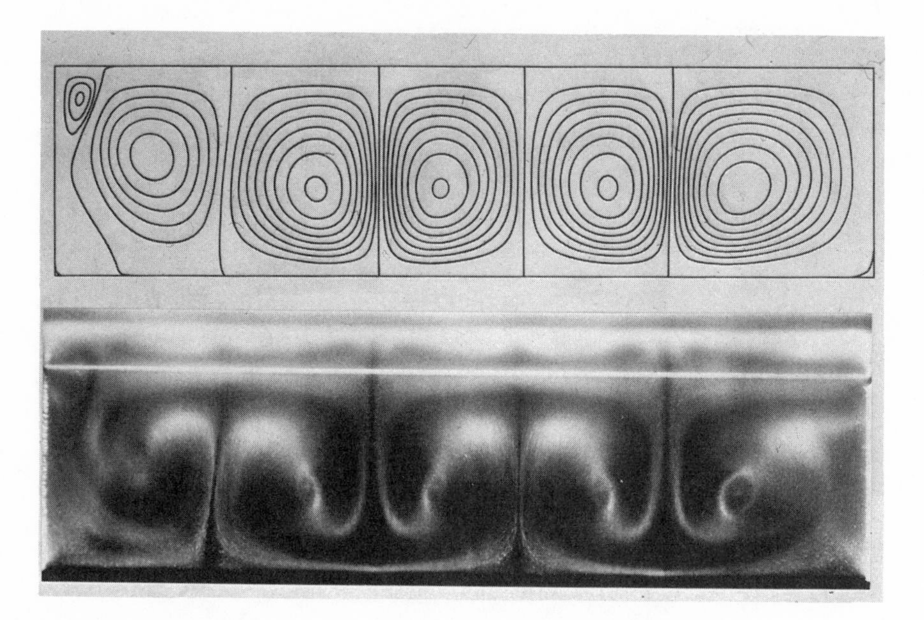

Figure 5.3 A cross section of the circular cells in a Taylor-Couette system compared with the same streamlines calculated on a computer.

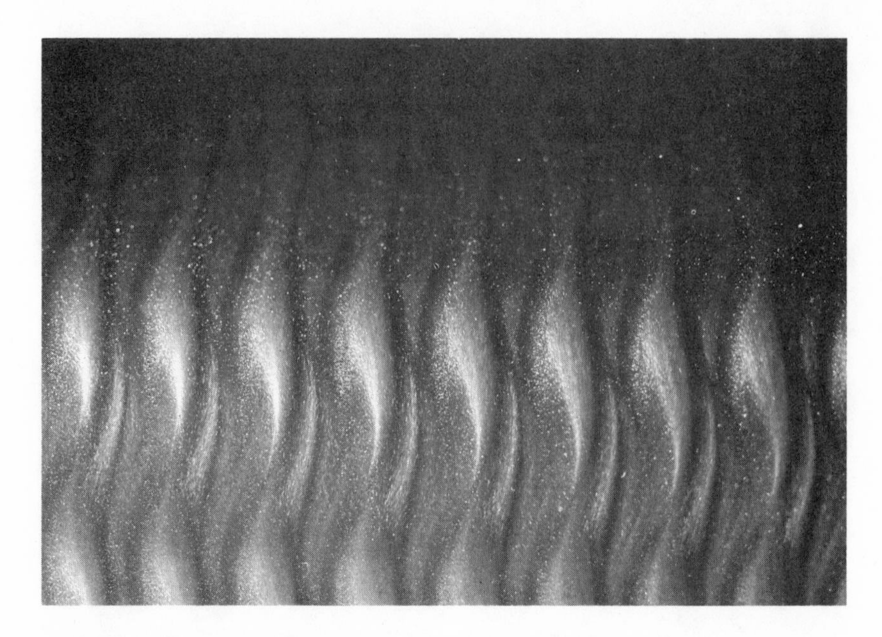

Figure 5.4 As a fluid flows faster in the Taylor-Couette system, waves start to appear as shown in this front view.

Figure 5.5 A further increase in the flow rate causes turbulence. Although the fluid flow is disordered in time, it is ordered in space.

appearing at diminishing intervals of *R*. Thus, *R* would rapidly reach a value where all possible frequencies arose in the flow. According to Landau, this is turbulence. Figure 5.5 of 'turbulent' flow in a Taylor-Couette system shows that although the fluid flow is disordered in time, there is still order in space. So it is very different from, say, a fast-flowing river, where the flow of water is completely disordered in both time and space.

Landau's description of turbulence is now being challenged. Two mathematicians, David Ruelle at Bures-sur-Yvette, near Paris, and Floris Takens at the University of Amsterdam, put forward an alternative view. They were studying sets of equations in the abstract, and suggested that chaos would ensue after only a few frequencies had appeared in the motion. To understand their point we need to consider a way of representing the data in terms of the 'phase portraits' described in Chapter 4.

The traditional way of analysing how a variable quantity, such as velocity, varies periodically with time – a time series – is to break down the overall complex signal into a series of sine waves of differing frequencies. This is called Fourier analysis. Takens suggested an alternative way of portraying the time series as a two- or three-dimensional map in 'phase space' (see Chapter 4). In its simplest form, you plot the time series against a time-delayed version of itself. A sine wave would give a circle in two-dimensional phase space. The time taken for the trajectory to rotate around the loop would then be the period of the oscillation. When there are two frequencies that are not related to each other, we need a second delayed time coordinate and another dimension to show it. We then obtain a torus, or doughnut. David Broomhead, Robin Jones and Greg King at the Royal Signals and Radar Establishment at Malvern have developed a powerful practical method which we have used to construct the results presented here.

If we disturb the trajectories in our phase space (this would correspond to perturbing the rotation of the inner cylinder, say) then a short transient phenomenon appears before the cyclic motion on the circle or torus is regained. We call these objects attractors in phase space.

Ruelle and Takens suggested that, instead of the appearance of a third wave as suggested by the Landau picture, chaos would arise through the emergence of a 'strange attractor' in phase space. We can consider the strange attractor as a region of phase space that attracts nearby trajectories, but inside the region neighbouring trajectories diverge and are chaotic in form. Harry Swinney and Jerry Gollub at the University of Texas at Austin have carried out experiments on the Taylor-Couette problem that support this picture.

Is there any more evidence to show the connection between observations in this fluid dynamical system and the chaos found in the solutions of simple sets of equations? An essential feature of any nonlinear system is that many different states can exist under the same operating conditions. In Taylor-Couette flow, these states would have different numbers of cells or waves or both. The state that you obtain depends on the history of its creation. Thus, if the system is large, there may be literally thousands of different states available – all of which can interact with each other. Controlling the experiment and interpreting the outcome becomes impossible, in any practical sense. We have to be vigilant when extracting qualitative features because the picture could be confused by the presence of many competing attractors.

One way to cope with this situation is to restrict the number of available states by limiting the physical size of the system. Gerd Pfister at the University of Kiel in West Germany experimented on the onset of chaos in a miniature version of the Taylor-Couette system containing a single cell. He uncovered a period-doubling route of the type already described in Chapter 3. This suggests that chaos plays an important role in the understanding of the results of this restricted flow.

If we now consider the sequence of events leading to chaos when the system is made a little larger, then new types of behaviour happen, each of which appears to be linked directly with modern thinking on chaos. The results from one of these studies are presented in the phase portraits in Plates 10 and 11. Plate 10 shows motion on a torus in our reconstructed 'phase space' and corresponds to the presence of two waves in the flow with very different timescales. It has qualitatively the same kind of structure as the output from some simple ordinary differential equations that you can solve on a microcomputer. If we change the control parameters of the experiment just a little, then chaos appears as shown in Plate 11. The 'shell' of the structure begins to fill in and the trajectories join the core at irregular intervals. However, the overall structure of the portrait is maintained and so we appear to have behaviour of the strange-attractor kind similar to that outlined above.

There does, therefore, appear to be a connection between observations of the onset of chaos in this fluid dynamical system and the onset of chaos in much simpler sets of equations. What is more, the chaos appears to arise through mechanisms that researchers have found in other situations such as chemical oscillators and lasers. You should remember, however, that the results are concerned only with the evolution of disorder in time in a very special flow: do not think that the challenging and exciting problem of turbulence of the type shown in Figure 5.6

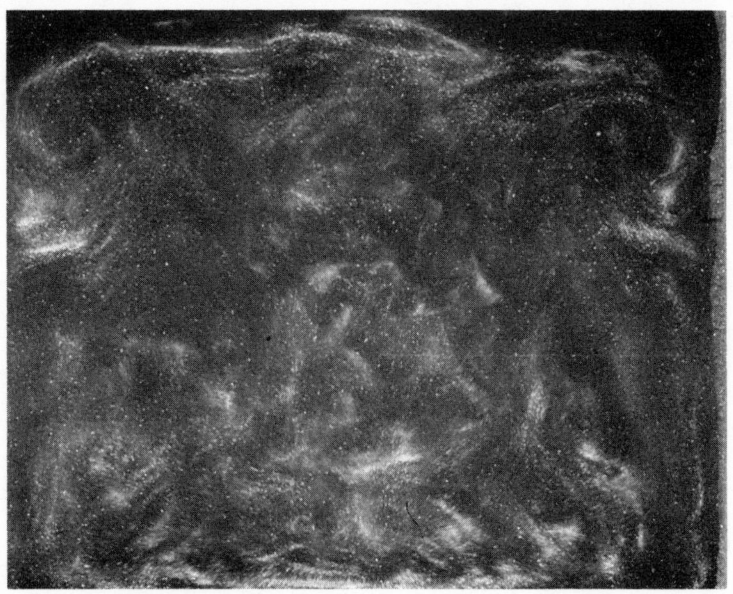

Figure 5.6 'Real' turbulence in a system where the fluid flows between a rotating cylinder and a square outer shell.

is solved. Nevertheless, mathematical ideas of chaos may have found a chink in the armour of this intriguing problem.

Further reading

D. J. TRITTON, *Physical Fluid Dynamics*, Oxford Science Publications, 1988.

6
A weather eye on unpredictability

TIM PALMER

Much of chaos theory came from trying to understand how the Earth's atmosphere behaves. Now, meteorologists are using chaos to assess how reliable climate and weather forecasts are.

Every day, meteorologists try to predict next week's weather using immensely complicated mathematical descriptions of how the atmosphere behaves. Research is also under way to develop models that will help them make predictions on an even longer timescale, seasonal forecasts of monsoon rains, for example. Meteorologists would even like to be able to estimate changes in climate resulting from human activities, such as the greenhouse effect.

And yet we know that the atmosphere is a chaotic system. As such, it is inherently unpredictable (see 'Chaos and high winds', below). So, are these attempts at longer-range weather and climate prediction a waste of time? Should we content ourselves with the television forecast of tomorrow's weather, and leave the rest to chance?

Although the weather can change every day as individual systems (for example depressions and their associated weather fronts) progress eastwards, certain spells of weather can last for weeks, months or even whole seasons. These spells are not characterized by individual weather systems, but by the position of the so-called jet streams, regions of strong wind in the upper atmosphere. They determine whether we will have a wet or dry summer, a mild or severe winter.

Chaos and high winds across the planet

The Sun's energy and Newton's laws are responsible for chaos in the atmosphere. Sunlight warms the tropical regions of the Earth, whilst over the poles, heat energy is radiated back to space, and there the atmosphere cools. The atmosphere attempts to pump heat from equator to pole in as thermodynamically efficient a manner as possible.

If the Earth were not rotating about its axis, the most efficient way of transporting this heat would be as follows: air heated at the Earth's surface in the tropics would rise in tropical latitudes, flow towards the poles in the upper atmosphere, sink over the poles, and return to the tropics in the lower atmosphere. This overturning motion across the planet would happen at all longitudes. In practice, it happens only in tropical and subtropical latitudes (with the large-scale sinking of air in the subtropics, giving rise to the desert regions of the Earth). In middle and high latitudes, the effects of the Earth's rotation on the dynamics of the atmosphere become important, giving rise to the so-called Coriolis effect, whereby projectiles tend to curve to the right in the northern hemisphere relative to an observer on the ground. As a result of the Coriolis effect, it is thermodynamically efficient to flux heat energy towards the poles through the depressions and anticyclones that we call weather systems.

Mathematically, we can describe these weather systems as fluid-dynamical instabilities of a spherical shell of rotating fluid, which is heated over the equator, and cooled over the poles. Theory correctly predicts the scale of these instabilities, about 1000 kilometres, with rates of growth corresponding to an amplitude doubling in a couple of days or so. The weather systems also flux momentum into middle latitudes from the tropics. This flux of momentum maintains the prevailing westerly winds against frictional dissipation at the Earth's surface. The westerly winds increase with height to a maximum in the so-called jet streams. The positions of the jet streams are not fixed in space and time, but meander, over distances of about 10 000 kilometres, longer length-scales than those associated with individual weather systems (see Figure 6.1). The positions of the jet streams can be used to define large-scale weather regimes; those determining the general pattern of weather experienced, say, during a period of a week or more, settled or unsettled, for example.

Although individual weather systems are instabilities of the larger-scale flow, they play a major role in maintaining the meanders of the jet stream on a planetary scale. From a mathematical point of view, this feedback between individual weather system instabilities and the larger-scale flow is a nonlinear process. Forced by the heat energy of the Sun, the atmosphere is, therefore, unstable and nonlinear. These two characteristics are the crucial components for chaos.

The evolution of a chaotic system is sensitive to the precise specification of the initial state; this means that irrespective of how complex our models become, or how accurate our weather data are, the laws of science impose a limit beyond which prediction of the weather is impossible.

Figure 6.1 shows the track of the jet stream over the Atlantic and Europe, which is associated with two different weather 'regimes'. In Figure 6.1a, the jet stream is more or less oriented along a line of latitude. Individual weather systems tend to be steered along the jet stream. Over the British Isles, this configuration would probably give a rather wet and unsettled spell of weather, as rain bands pass by with monotonous regularity.

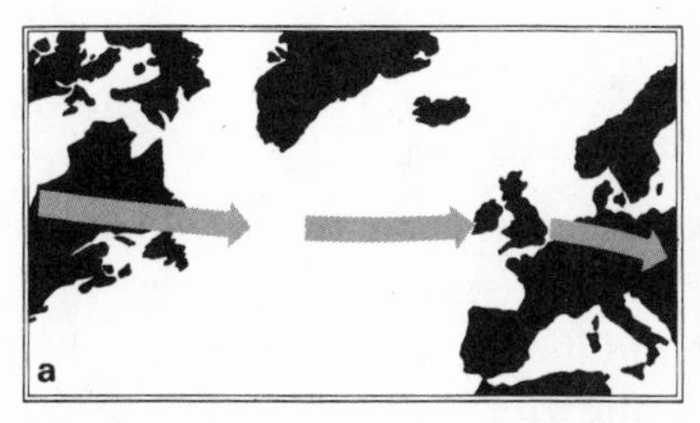

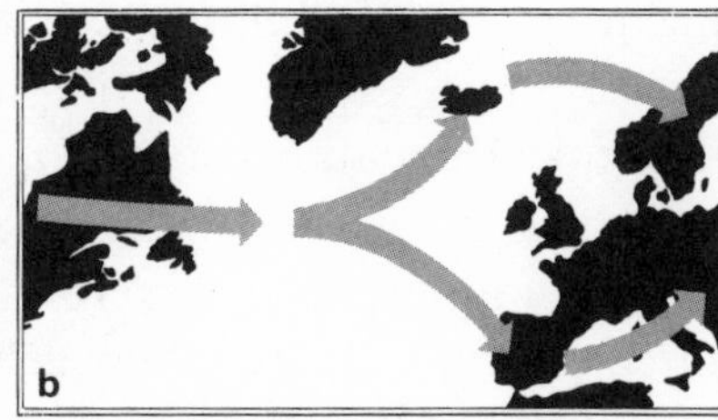

Figure 6.1 The position of the jet stream over Europe and the Atlantic in two weather regimes. The first gives rather unsettled weather over the British Isles; the second may give fine warm weather in the summer, dull and possibly severely cold weather in the winter.

Meteorologists call this a 'zonal' regime. In Figure 6.1b, the jet stream splits into two branches over the mid-Atlantic, with one branch positioned north of the British Isles, the other to the south. In the summer, this configuration brings about a warm fine spell of weather over Britain; in the winter, it produces dull, overcast, and possibly very cold weather. Meteorologists usually refer to this as a 'blocked' regime.

We can define weather regimes quantitatively from historical records of data of weather in the northern hemisphere, using what are called 'cluster-analysis' techniques. In practice, about 10 different regimes characterize most of the large-scale variability of the atmosphere in the northern hemisphere.

Meteorologists have long been interested in the predictability of these weather regimes. Can we forecast how they evolve up to a month ahead, even though we can predict what happens to individual weather systems for only a few days ahead?

These sort of problems motivated the meteorologist Edward Lorenz, whose work at the Massachusetts Institute of Technology in the early 1960s spawned much of the activity in chaos theory today. The atmosphere behaves like a turbulent fluid, and Lorenz was only too aware that it was governed by a set of mathematical equations that were nonlinear and were extremely sensitive to small changes – in other words, showed modes of instability. He had an intuitive feeling that this would make weather prediction a tough problem. To confirm his hunch, he sought a way of simplifying these equations so that he could study them mathematically, while retaining their essential nonlinearity and instability.

The most drastically simplified version of the full fluid-dynamical equations led to a 'model climate' with just three variables, x, y and z. A state of instantaneous 'weather' in Lorenz's model can, therefore, be represented by a point in a three-dimensional 'phase space', and the evolution of the weather with time can be represented by a line, or trajectory, in this space (as described by Ian Stewart in Chapter 4). The climate of the model, the set of all possible model weather states, is known as the Lorenz attractor (see Figure 6.2).

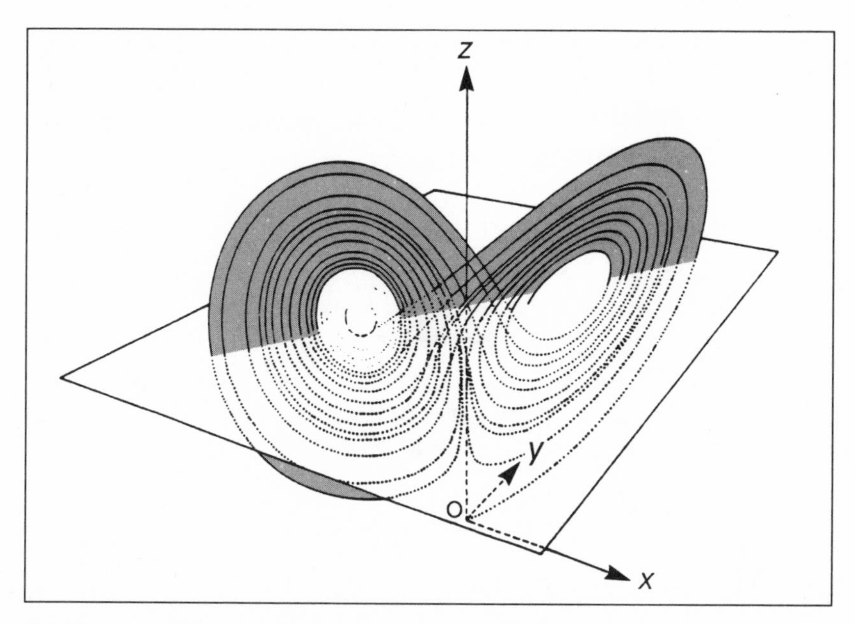

Figure 6.2 The Lorenz attractor in the three-dimensional phase space spanned by the Lorenz model variables x, y, z.

This attractor has no volume in this three-dimensional space, yet is neither a simple one-dimensional line, nor a smooth two-dimensional surface. The attractor has a fractional dimension (2·06), and therefore, not surprisingly, carries the epithet 'strange' (see Chapter 4). It represents one of a generic class of strange attractors whose topology characterizes the chaotic, unpredictable properties of the basic equations.

Although the three-component equations that Lorenz proposed do not realistically describe the evolution of weather regimes, they have similar chaotic properties to more realistic models. So we can use the Lorenz model to describe in a qualitative way the chaotic behaviour of the evolution of weather regimes in the atmosphere.

First, notice from Figure 6.2 that the Lorenz attractor has two separate branches, sometimes called butterfly wings. We can think of these wings as representing, in our abstract state space, the two weather regimes shown in Figure 6.1 in real space. For the sake of argument, suppose the left-hand wing corresponds to the zonal regime of Figure 6.1a, and the right-hand wing corresponds to the blocked regime of Figure 6.1b. In other words, any two points on the left-hand wing relate to different instantaneous weather, but the large-scale flow would be the same.

Imagine two points arbitrarily close to each other, on the left-hand wing of the attractor. Using our conventions, these two points represent almost identical weather states in a regime characterized by rather unsettled weather conditions over the British Isles. We now follow the initial evolution of these two weather states. There are three possibilities: both trajectories remain on the left-hand wing (see Figure 6.3a); both trajectories evolve towards the right-hand wing, as in Figure 6.3b; or one trajectory remains on the left-hand wing, while the other moves to the right-hand wing, as in Figure 6.3c. Note that in all three cases, the two trajectories have diverged, implying quite different forecasts of instantaneous weather. On the other hand, in the first two scenarios (see Figures 6.3a and 6.3b) the two trajectories evolve to the same weather regime (remaining unsettled in the first case; becoming more settled in the second case).

You can see, therefore, that although the atmosphere is fundamentally chaotic, you can predict the weather regime from

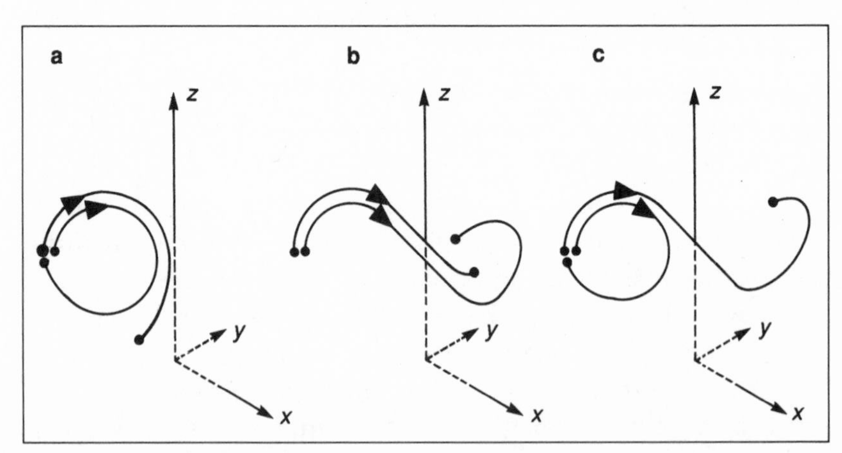

Figure 6.3 Three trajectories showing the evolution in Lorenz's model atmosphere from almost identical weather states. In **a** and **b**, the states evolve similarly; in **c**, they evolve differently.

certain initial conditions in the atmosphere. To find out what these initial conditions are, we need to make a series of weather forecasts from (a sufficiently large number of) similar but not identical initial states. Figure 6.4 shows the evolution in phase space of two ensembles of realistic weather forecasts over a certain period. For the first set of initial conditions, in Figure 6.4a, the forecasts start to diverge only a little, indicating that the predictability for that set of initial conditions in the atmosphere is pretty good, so we can confidently forecast the evolution of the weather regime over that period.

On the other hand, in Figure 6.4b, the paths of evolving weather patterns from similar initial conditions disperse considerably, indicating that the atmosphere is in a particularly chaotic state during the period of the forecasts, so we cannot make any meaningful predictions.

Now, the Lorenz model has only three variables, or three degrees of freedom. It is much too simple to describe accurately the evolution of the real atmosphere. In fact, increasing the number of degrees of freedom in models has improved the quality of weather prediction for the first few days ahead. Today's weather-prediction models have about a million degrees of freedom.

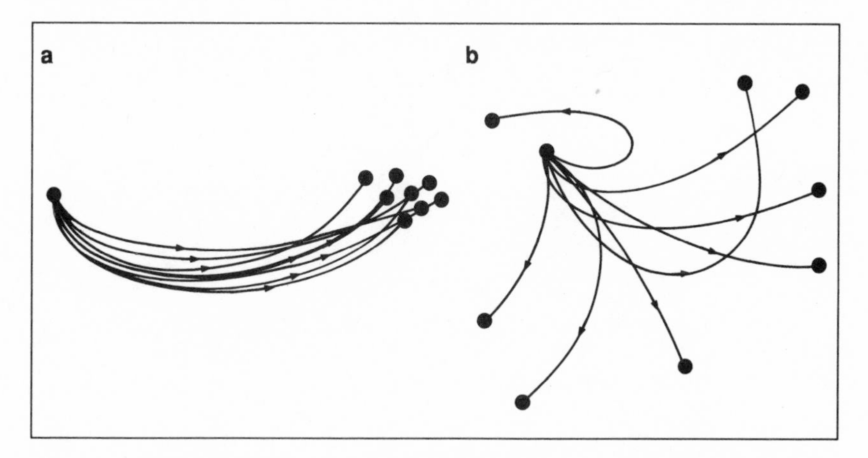

Figure 6.4 The evolution of an ensemble of forecast trajectories in the phase space of a realistic model for weather prediction. In **a**, we can give the forecast with confidence, but not so in **b**.

Figure 6.5 Initial conditions for eight forecasts of the weather-prediction model of the European Centre for Medium-Range Weather Forecasts. Arrows show the direction of the wind.

Figures 6.5 and 6.6 show an example of an ensemble of eight forecasts from the weather-prediction model we use at the European Centre for Medium-Range Weather Forecasts (ECMWF). The pictures show contours of the height of a

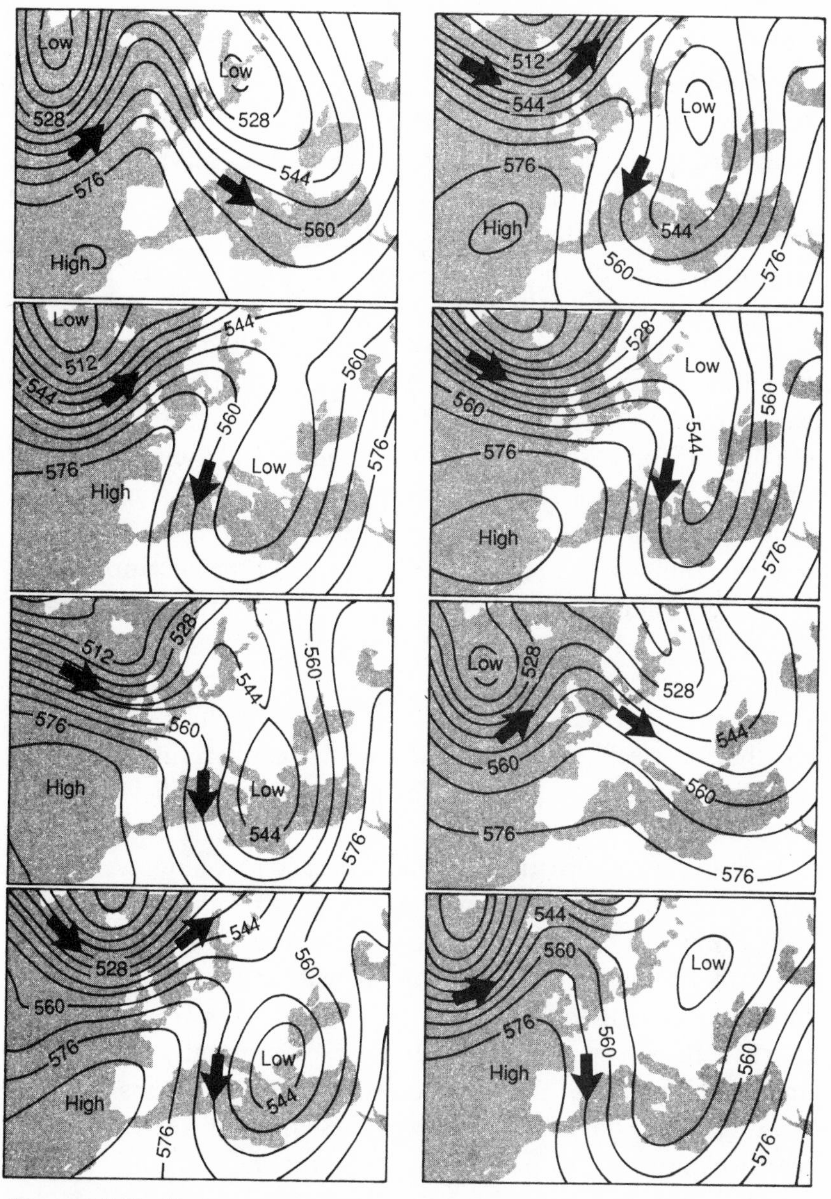

Figure 6.6 Predictions one week later for the eight forecasts. The ensemble shows that we can predict the weather for Italy with more confidence than we can for Britain.

pressure surface (500 hectopascals) in the middle of the atmosphere. The wind blows parallel to these contours in the direction shown by the arrows – the strength of the wind is proportional to the contour gradient. Figure 6.5 shows the initial conditions for the eight forecasts. Note that they look very similar. Differences between individual members of the ensemble correspond to uncertainties associated with the weather observing network. The initial flow in Figure 6.5 corresponds approximately to the pattern shown in Figure 6.1; over Britain, the flow is very weak, with the region of strong winds splitting over the east Atlantic towards the north-east and south-east.

We now run the computer model eight times from each of these initial conditions and see what the model predicts for the weather one week later. This is shown in Figure 6.6. You can see that the forecasts are now far from identical; their trajectories in phase space (not shown) have begun to diverge significantly. However, over some parts of Europe, you can see a weather pattern common to many of the members of the ensemble. Over Italy, for example, the initial state showed an air flow from the south; after a week, the winds swing round so that they have a northerly component. So, an Italian forecaster could confidently predict that Italy should expect much colder weather a week ahead. Using the ensemble, the forecaster could estimate probabilities that temperatures will fall within certain ranges.

By contrast, the British weather forecaster would have a more difficult task – the forecasts over Britain diverge much more (confirming the saying about the unreliability of British weather). Nevertheless, it would not be unreasonable to predict a trend to more westerly flow and unsettled weather. This example is one that is neither particularly chaotic, nor exceptionally predictable, and highlights the practical consequences of chaos for the weather forecaster. There are many research centres around the world exploring the possibility of predicting the evolution of weather regimes up to a month ahead. It is early days in this business and meteorologists still need to refine their description of various important physical processes in the computer models. If they are to forecast weather regimes this far ahead, they will have to run ensemble forecasts of the sort just described, so as to be able to estimate how reliable the prediction is. At present, raw computing power is still insufficient to do this routinely. But with advances in computer technology, this kind of forecasting may not be many years away.

So far, we have talked about weather in the middle latitudes. In the tropics, the dynamics of the atmosphere is somewhat different, mainly because the Coriolis effect of the Earth's rotation is less dominant. In particular, although there are weather systems in the tropics, such as hurricanes, monsoon depressions and so on, that arise from instabilities of the larger-scale flow, these weather systems do not feed back into the larger-scale flow to the same extent as weather systems in the middle latitudes. In fact, the behaviour of the large-scale tropical atmosphere is very strongly linked to the temperature of the ocean surface, which evolves over months, rather than days. One well-studied phenomenon that intimately involves the coupled dynamics of the ocean-atmosphere system is the so-called El Niño event, during which ocean temperatures in the tropical east Pacific can rise up to 4°C above normal during a season. Meteorologists believe that El Niño can influence weather patterns over a substantial fraction of the globe.

It may be feasible to predict flow on a planetary scale in the tropical atmosphere for a season, using models that take into account the dynamics of both the atmosphere and the oceans. Certainly there have been some encouraging successes in forecasting El Niño and its consequences a season ahead. In a few years' time, weather forecasters may be able to predict seasonal rain over Africa, India and other tropical countries from such weather prediction models. However, nonlinearity and instability, the hallmarks of chaos, are not totally absent in these tropical predictions, and it is likely that the 'multiple-ensemble integrations' which people have already applied to weather at mid-latitudes will still be a necessary tool for the tropical forecaster.

Finally, does chaos theory prevent us from predicting possible climate change in the next century? The answer here is no. The type of prediction is quite different from that outlined above. Here, the goal is not predicting an individual trajectory on the climate attractor; the goal is to determine the shape and position of the whole climate attractor itself when, for example, greenhouse gases increase.

The critical question that climatologists are trying to answer is whether the climate attractor will suffer a minor perturbation (for example, a small shift of the whole attractor along one of the axes of phase space), or whether there will be a substantial change in the whole shape and position of the attractor, leading to some possibly devastating weather states not experienced in today's climate. Chaos theory certainly does not forbid the possibility of some substantial change to the atmosphere's climate attractor as a result of modest increases in the amount of carbon dioxide. At the moment, we cannot be sure of the answer. The same sort of models employed in predicting the weather are also being used to try to find out what the greenhouse effect has in store for us. As with attempts at predicting weather regimes, uncertainty in climate forecasting is mitigated by the fact that several of the climate centres around the world now have sophisticated models. Again, researchers can evaluate their confidence in predicting greenhouse warming in terms of the dispersion of the ensemble of predictions from these different models.

Nevertheless, chaos dynamics should caution us from making too premature a judgement about climatic changes. Returning to the Lorenz attractor, there is no pre-ordained number of times that a given trajectory must circle around one of the butterfly wings; it could be once, 10 times, or 100 times, depending on the precise position of the trajectory on the attractor. If we have a mild winter, a warm summer, then another mild winter, we might not necessarily be in the throes of man-made climate change; the system might just be revolving happily around one small part of phase space, and on that hundred and first revolution, the system might, quite unexpectedly, and for no apparent reason, evolve towards another part of phase space, which is associated with cold winters and miserable summers. For this reason, many meteorologists are quite guarded about whether global warming due to the greenhouse effect really has arrived.

To the lay person, the unpredictability of the weather may

be a curse; to the meteorologist, it is what makes the subject fascinating, and fun to study. Above all, chaos does not mean that we must throw in the towel, and leave all to chance. This blend of fundamental science with state-of-the-art computer technology is leading to unparalleled insights into the workings of the fragile gaseous envelope that surrounds and sustains us.

Further reading

JOHN MASON, PETER MATHEWS and J. H. WESTACOTT (eds.), *Predictability in Science and Society*, Cambridge University Press, 1984.

H. GHIL and S. CHILDRESS, *Topics in Geophysical Fluid Dynamics: Atmosphere Dynamics, Dynamo Theory, and Climate Dynamics*, Applied Mathematical Sciences, Vol. 64, Springer Verlag, 1987.

7
The chaotic rhythms of life

ROBERT MAY

In the 1970s, population biologists helped to launch the theory of chaos. Now it seems that many aspects of life are probably chaotic. But the problem is that they are also difficult to study.

One of the most memorable museum exhibits I have ever seen is in the Smithsonian Museum of Natural History in Washington D C, where the floor, cupboards and ceiling of a kitchen are covered with the thousands of cockroaches that would be produced by an average female cockroach if all her offspring survived. The exhibit vividly illustrates one of the basic tenets of Darwinian evolution: all animals have the capacity to do more than replace themselves. But most of the time, a variety of factors – predators, limited food supplies, disease and a myriad others – hold the populations in check.

The result is that most populations of plants and animals usually fluctuate. They tend to increase after dropping to unusually low densities (at which point, the conditions become most suitable for maximum growth) and, after reaching unusually high densities, they tend to decrease again. One of the main aims of ecologists is to discover just what the 'density-dependent' effects regulating populations are. Such understanding is not only fundamentally important, but it also has practical applications in trying to predict the likely effects of natural or man-made changes such as occur when a population is harvested or when climate patterns alter.

Until recently, most ecologists assumed that the effects regulating density would, in the absence of other factors, keep a population at some constant level, and that the irregular fluctuations actually seen in so many natural populations resulted from unpredictable ups and downs in various environmental influences. So ecologists studying population dynamics saw their task as trying to extract a steady signal from the masking overlay of environmental noise.

But in the early 1970s, George Oster at the University of California at Berkeley, Jim Yorke at the University of Maryland, I and others began to look more closely at the equations that fish biologists and entomologists had proposed to describe fluctuations in populations. We found that these equations show an extraordinary variety of dynamical behaviour, surprisingly richer than biologists had previously assumed.

Take the equation $x_{t+1} = \lambda x_t(1 - x_t)$, which can describe how a population behaves, and whose rich mathematical character Franco Vivaldi has already discussed in Chapter 3. Here x_t may represent the population of an insect, with the subscript t labelling each successive, discrete generation. Suppose that each adult in generation t would produce λ offspring if there were no overcrowding. Then the population of the next generation, x_{t+1}, would be λx_t. The additional factor $(1 - x_t)$ in the equation represents the feedback from effects due to density or crowding. The population density is scaled such that beyond a crowding level $x = 1$, it goes to zero (that is, negative values of x correspond to extinction).

Beginning in the early 1970s, these ecological studies have brought this important equation to centre stage in many scientific disciplines. When λ is less than 1, the population decreases to zero (for the obvious reason that its reproductive rate is below unity). When λ is greater than 1, but less than 3, the population settles to the constant value that intuition would suggest. Further increases in λ result in an increasing propensity for the population to 'boom' when its density is low, and 'bust' when it is high.

This increasing tendency to 'boom and bust' shows up as positive feedback (akin to the shrieks from a microphone when the power level is turned too high) and the population oscillates in a cycle with a period of two generations, alternating between a high and a low value. As λ continues to increase beyond 3, these cycles become more complex, with the period lengthening, under successive doublings, to four generations, then eight, then 16, and so on. The population continues, however, to alternate between high and low in successive generations. Finally, when λ is bigger than about 3·57, a domain of apparently random fluctuations appears. This is chaos. The simple iterative rule now generates population values that look for all the world like samples from some random process. Figure 7.1 shows the spectrum of possibilities.

For population biologists, the first message from all this is that the signals from the purely deterministic processes controlling the population density can look like random noise. Even more disconcerting is that, as previous chapters have described, in a chaotic system, although the starting values of x might be quite close together, they diverge fairly rapidly, eventually leading to quite different trajectories (see Figure 7.2). This sensitivity to initial conditions means that long-term prediction is impossible.

Why didn't people recognize earlier these properties of what is a very simple equation? Several mathematicians had unravelled

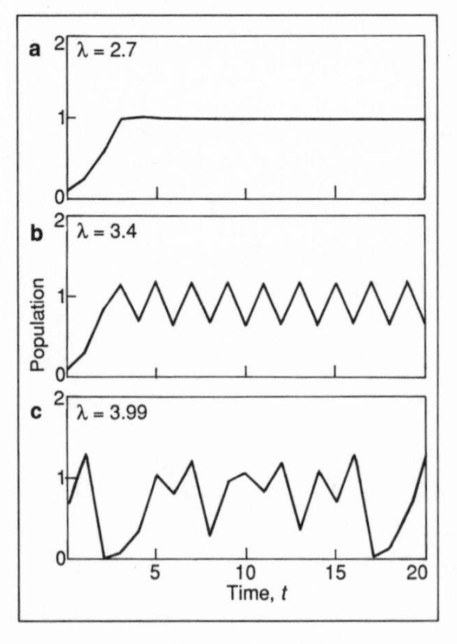

Figure 7.1 Different values of λ in our simple equation produce widely differing patterns. In **a**, the population settles to a constant level; in **b**, the population alternates between high and low in successive populations; and in **c**, the population behaves chaotically.

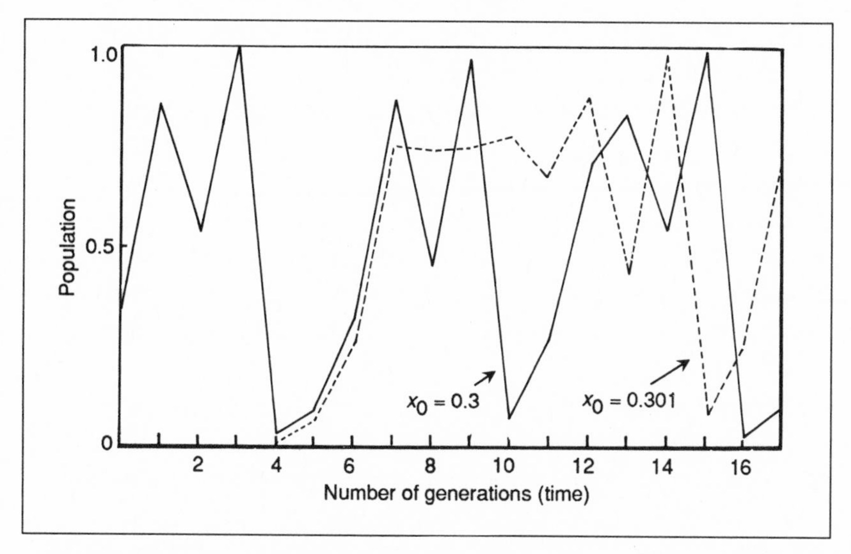

Figure 7.2 shows how sensitive the growth of a population is to the initial conditions when the dynamics are chaotic. Although the initial values for two populations differ only by 0·3 per cent, their trajectories rapidly diverge.

its mathematical characteristics, but failed to realize what they implied for the real world. On the other hand, several ecologists, such as William Ricker who worked on fisheries at the University of British Columbia and P. A. P. Moran who worked on insect populations at the Australian National University in Canberra, had actually studied the equation. They were seeking steady solutions, however, and having found them they conveniently forgot the chaotic behaviour that they had also noticed. What happened in the early 1970s is that ecologists with sufficient mathematical know-how to understand the equations looked at them in practical settings, so grasping their wider implications.

To study the dynamics of a population in the light of such deterministic equations, we must bring our sample species into the laboratory. Here, we can control the environment and eliminate the complicating effects of interactions with other populations. The result is a kind of living computer – useful, but not giving a reliable picture of how the population really behaves in nature, where other species or environmental changes may strongly affect the dynamics.

There have been a few such laboratory studies, using quite small creatures, such as rotifers, *Daphnia* and blowflies, which have the advantage that they do not take up too much space and their generations tick over fairly fast. By raising the temperature, for instance, the experimenter can speed up metabolic processes so that the fluctuations in population become more pronouncedly 'booming and busty'. Such studies have, indeed, behaved as expected from the equation. But they do not produce the beautifully crisp period doublings and other phenomena that make the corresponding experiments on physical systems so compelling.

In the natural world, the job of filtering the information that we want – the density-dependent signals – from superimposed environmental noise is hard enough if the underlying dynamics are steady. But if the signal itself is chaotic, the situation is even more complicated. One powerful method for exploring this problem is to create a computer model, which generates sets of artificial 'pseudo-data' representing the size of a population, generation by generation.

In this imaginary world, the investigator can specify all the factors governing the population's size. The researchers can then analyse these pseudo-data using the methods normally applied to real data from the field. In this way, they can judge whether the methods do indeed lay bare the mechanisms controlling a population's density that were built into the imaginary world.

Michael Hassell of Imperial College, London, for example, employs rules that encapsulate his ideas about the factors influencing certain insect populations. In his computer models, adult insects are distributed, according to rules with some random elements, in many patches, say on leaves, twigs or bushes. These adults lay eggs. The probability of ensuing larvae surviving depends on how crowded they are in each patch; this is the essential density-dependent factor. Surviving larvae mature into the next generation of adults, who then spread out into other patches and begin the whole process anew.

Figure 7.3a shows one set of such pseudo-data for the overall population, generation by generation. The fluctuations come

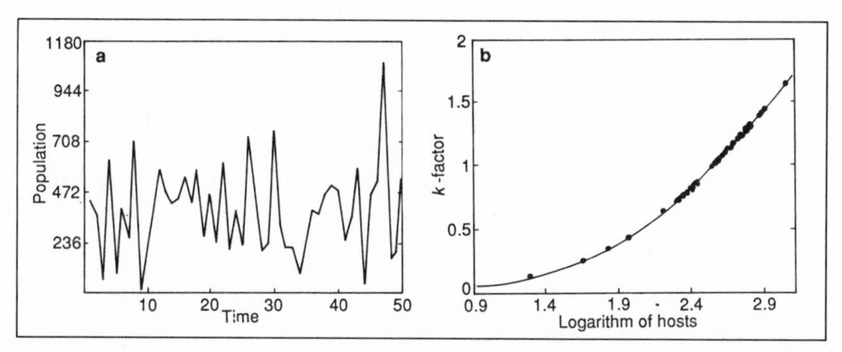

Figure 7.3a shows the size of an imaginary insect population, year by year, as given by a computer model. The variations are due to a mixture of chaotic dynamics and real noise. *Figure* 7.3b gives the curve through the points indicating that conventional *k*-factor analysis can unscramble the signal from the noise, and so detect the effects of insect-population density on the dynamics.

partly from random elements in the process of dispersal (mimicking a natural situation), and partly as a result of deterministic chaos from density-dependent effects in crowded patches. Figure 7.3b shows that analysing the data, with a conventional procedure called '*k*-factor' analysis (which aims to reconstruct the underlying map relating the sizes of populations in successive generations), gives a simple curve that uncovers how the survival of the larvae does depend on their density in each patch.

Now, Figure 7.4a repeats this exercise, but here we introduce environmental randomness into the model by varying the

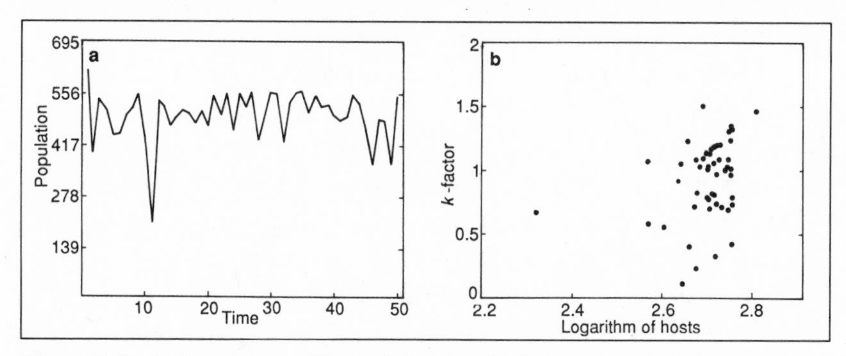

Figure 7.4a is the same as Figure 7.3a, but includes a random factor in the number of eggs laid by an adult insect. Again there is a mixture of chaotic and genuinely random effects. *Figure* 7.4b This time, *k*-factor analysis just gives a cloud of points. That is, the standard methods of analysis fail to expose the rules governing this imaginary world.

number of eggs laid by each adult (again, the size of these fluctuations accords with real examples). As before, the fluctuations in population sizes from generation to generation come partly from these random factors and partly from deterministic chaos. But applying the same conventional methods to these pseudo-data reveals no discernible signal, as shown in Figure 7.4b. So, although Hassell knows exactly what governs his world (because he constructed it), the standard techniques fail to show what is really going on. The basic problem is that once patchiness in distribution, environmental noise, and chaotic dynamics all interweave, it can be difficult to disentangle the chaotic, density-dependent signals from additive noise.

There is, unfortunately, no punch line to this part of the story. In the field, chaotic dynamics can create difficulties that we do not fully understand, and which may require more detailed studies than have been usual in the past.

Another complication, which Oster and I pointed out in 1976, is that the population we are studying usually interacts with other species, which in turn interact with others, creating a sort of biological many-body problem. Such webs of interactions make chaotic dynamics much more likely. It means that we have to add an extra variable for each species included in the study, resulting in a multidimensional system of equations of the kind described in previous chapters. So understanding population dynamics then becomes a formidable task.

In these circumstances, William Schaffer at the University of Arizona, Mark Kot at the University of Tennessee, George Sugihara at the Scripps Institution of Oceanography in San Diego, and others, have explored ways of analysing data to see whether the dynamics are truly chaotic. One approach is to look for mathematical indicators of chaos underlying the dynamics, such as a 'strange attractor' in multidimensional phase space, as described by Ian Stewart. We can then try to construct the attractor – without any understanding of the fundamental biological mechanisms generating it – and set it in the appropriate dimensions to make it 'come into focus'.

Taking this approach, Schaffer and Kot looked at cases of

measles in New York City over a 40-year period. They found that, for example, the string of monthly data, or time series, for the numbers of measles cases from 1928 to 1968 – when vaccination began to alter the dynamics of the system – revealed a three-dimensional attractor. So-called 'Poincaré cross-sections', or planes slicing across the attractor, suggest that the dynamics correspond to deterministic chaos generated by an approximately one-dimensional map, or an equation of the kind we have already described.

Researchers have analysed these, and other measles data from Copenhagen, using other methods of detecting chaos. In all cases, the conclusion is that deterministic chaos best explains the data, although the length of the time series (at best some 500 monthly points for the New York data) is too short for a truly reliable analysis by these data-hungry techniques.

This approach of distinguishing between chaos and random noise in population biology is in its infancy. Even when successful, such methods tell us only that there are some nonlinear, density-dependent mechanisms operating, but do not tell us what the mechanisms are. To some ecologists this has an air of black magic. But I think the approach is useful. It can show us when it may be profitable to search for such mechanisms and to attempt to make short-term predictions from apparently noisy data.

You might suspect that you could apply similar kinds of mathematical analysis to other areas of biology where feedback might lead to chaotic changes. You would be right. Chaos may be important for understanding some aspects of how genetic variability is maintained in natural populations.

You only have to look around you in the street to see that human beings differ a great deal, for example, in height, weight and facial appearance. How is this variability, or polymorphism, generated and maintained in a species? One way is through natural selection that depends on the relative or absolute abundance of individuals with the same genetic makeup, or genotype, in such a way as to favour rarer genotypes. There are many ecological effects that result in a rarer genotype enjoying a selective advantage. As J. B. S. Haldane first emphasized in 1949, the effects of infectious diseases are particularly important because diseases spread more effectively among more crowded populations. If different genotypes of hosts have differing degrees of resistance to different strains of a pathogen, then the rarest genotypes will enjoy a selective advantage. The reason is that the pathogens afflicting them will spread less effectively, or not at all, because the hosts are more spread out.

Until recently, conventional analyses of population genetics showed that such selective effects could maintain variability within a species, but these static analyses tended to assume that the proportions of the different genotypes remained constant over time. William Hamilton at Oxford, Simon Levin and David Pimentel at Cornell, Roy Anderson at Imperial College, London, and I have more recently studied the dynamic properties of the interactions among hosts and pathogens. The studies show that the proportions of any one genotype are likely to fluctuate chaotically from generation to generation. Such chaotically fluctuating polymorphisms are likely to be the rule rather than the exception.

So far, few biologists have investigated changes in the proportion of different genotypes present over time in real populations. Karen Forsythe of the Walter and Eliza Hall Institute of Medical Research in Melbourne has shown that the predominant strain of malaria in people in New Guinea differs from village to village, or in the same village over time, in ways that look chaotic. Studies of patchily distributed populations of plants in the Snowy Mountains by Jeremy Burdon of CSIRO in Canberra also give enigmatic hints of chaotic changes in gene frequency.

What is clear is that the selective mechanisms that maintain genetic diversity within populations can do so at chaotically fluctuating levels. There is currently much excitement about sequencing what is called *the* human genome. Evolutionary biologists have long recognized, however, that understanding variability within human genomes will be just as exciting. Chaos could add an extra dimension to this enterprise.

Given the number of biological and environmental factors

likely to influence the dynamics and genetics of natural populations, we might expect to find more unequivocal examples of chaotic dynamics at the sub-organismal level, in physiological or neurobiological processes.

Leon Glass and Michael Mackey at McGill University in Montreal were among the first to explore the possibility that many medical problems may be what they call 'dynamical diseases', produced by changes in physiological factors that cause normally rhythmic processes to show erratic or chaotic fluctuations. For instance, in some blood diseases the numbers of blood cells show large oscillations that are not normally present. Glass and Mackey showed that simple, but realistic, mathematical models for controlling blood cell production display the same periodic and chaotic oscillations as seen clinically when some parameter is varied. Such changes in the parameters of the model have a physiological interpretation. Clifford Gurney of the University of Chicago has performed experiments, based on Mackey and Glass's models, which produce oscillations in numbers of blood cells in mice.

Breakdowns in cardiac rhythms are obvious candidates for 'dynamical diseases'. The best studies of the dynamics of heartbeats, however, come from Petri dishes, not humans. Glass and his colleagues, Michael Guevara and Alvin Shrier, showed that a cluster of heart cells from chick embryos will beat spontaneously with an innate and regular rhythm (see Figure 7.5). Applying a strong electric field to this cell aggregate resets the phase of the heartbeat; that is, the next beat will be earlier or later than normal. Introducing a periodic series of such electrical impulses means that the heart is pushed by two forces with different periods: one with the heart cells' intrinsic rhythm and the other with the rhythm of the electrical shocks. The ensuing heartbeat depends on the relation between these two periods.

In some cases, the heart cells resonate with some harmonic of the stimulus, beating once for each jolt, or twice, or perhaps three times, for each two jolts, and so on. In other instances, the cells fire apparently at random, giving irregular or chaotic patterns. Glass and his colleagues interpret the dynamics of these

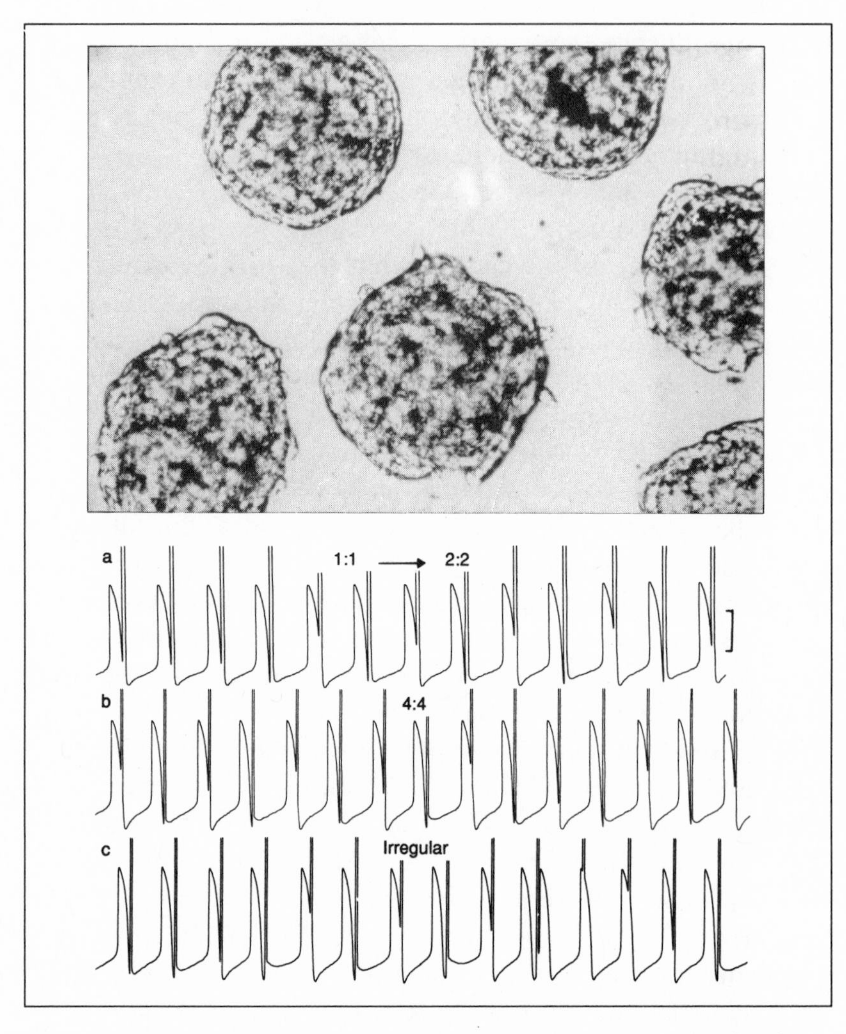

Figure 7.5 Periodic electric shocks applied to spontaneously beating heart cells from chick embryos (*top*) can cause the heartbeat to undergo period doubling, as in traces **a** and **b**. In trace **c**, the electrical stimulus produces a chaotic heartbeat.

periodic or chaotic patterns in terms of the complex bifurcations that result from the interplay between the innate physiological rhythms of heart cells and the frequency of the forcing electrical stimuli. These experiments show that you can induce and study chaos in an artificial system that is a metaphor for cardiac processes. Applying these to cardiac arrhythmias, or to electrocardiograms before and after heart attacks, is, however, still at an early stage.

Neurophysiology also offers a wide range of phenomena that are candidates for 'dynamical diseases', or abnormal oscillations and complex rhythms posing therapeutic problems. Sometimes, there is a marked oscillation in a neurological control system that does not normally have a rhythm. Examples are ankle tremor in patients with corticospinal tract disease, various movement disorders (Parkinson's tremors, for instance), and the abnormal paroxysmal oscillations in the discharge of neurons that occur in many seizures.

Alternatively, there can be qualitative changes in the oscillations within an already rhythmic process, as in abnormalities in walking, altered sleep–wake cycles, or rapidly cycling manic depression. Yet again, clinical events may recur in seemingly random fashion, as in seizures in adult epileptics. Neural processes are, however, so complex that it is not easy to see how models for these dynamical diseases – if, indeed, they exist – can be developed, tested and understood.

One approach, taken by Paul Rapp at the College of Medicine of the State University of Pennsylvania, rests on analysing the dynamical complexity of electroencephalograms (EEGs) which recorded the patterns of brain activity of human subjects as they performed various tasks. Rapp found that the complexity of the patterns changed in response to changes in intellectual effort. One study, for example, asked subjects to count backwards from 700 by sevens. Rapp characterized the changing complexity of the resulting EEG patterns using what is becoming a standard method for analysing chaotic rhythmic processes; he computed the 'fractal dimension' of the jagged time series, and found the dimension rose from its background value of around 2·3 to around 2·9 during the tests. (See Chapter 10 for a description of fractal dimensions.) He infers that the higher-dimensional, more complicated EEG patterns correspond to a more alert state.

If we want to understand the dynamics of neurophysiological processes more clearly, we need a simpler system that we can control, such as the light-reflex of the pupil of the eye. This reflex is a neural control mechanism with a delayed negative feedback, which regulates the amount of light reaching the retina by changing the area of the pupil. You can see the phenomenon informally by playing with a light in front of a mirror. If you want to publish in a scientific journal, however, you would do better to control the observations by 'clamping' the pupil light-reflex; the feedback loop is first 'opened' by focusing a light beam onto the centre of the pupil; the loop is then 'closed' by an electronically constructed circuit, or 'clamping box', which relates measured changes in pupil area to changes in the light shed on the retina.

The time delay in the pupil's response, or 'pupil latency', is around 0·3 seconds. As the gain and/or delay in the feedback loop increases, the pupil light-reflex becomes unstable and starts to oscillate periodically. In another experiment, where Andre Longtin, John Milton and colleagues (also at McGill) designed the clamping box to mimic 'mixed' feedback, the pupil reflex became unstable and produced aperiodic oscillations. This is due to the interaction of complex, possibly chaotic, dynamics and neural noise.

These physiological and neurological studies are reminiscent of those on single populations of animals, in that it is hard to apply theory to the real situation. Theory and experiment do agree, with varying degrees of precision, in the laboratory, but these simple, artificial demonstrations are always open to the cavil that they are no more than animated computer experiments. Many evolutionary biologists think that chaotic dynamics do not exist among real populations, because the accompanying fluctuations carry the risk that subpopulations will wink out, patch by patch, rendering long-term persistence unlikely. By the same token, earlier work tends to see chaos as a villain in physiology, manifesting itself in 'dynamically diseased' arrhythmias or seizures.

Ary Goldberger at Harvard Medical School has argued, to the contrary, that chaos gives the human body the flexibility to respond to different kinds of stimuli, and in particular that the rhythms of a healthy heart are chaotic. Goldberger bases his claims on analyses of electrocardiograms of normal individuals and heart-attack patients. He argues that healthy people have ECGs with complex irregularities, which vary systematically on timescales from seconds to days, whereas people about to experience a heart attack have much simpler heart rhythms. Critics correctly observe, however, that the broad patterns do not necessarily imply chaos, and that more emphasis should be put on studying the dynamics of heartbeats and the physical performance of the heart.

On an even more speculative note, Alisdair Houston at Oxford has pointed out that there is one context where chaotic unpredictability certainly could be useful. Organisms seeking to evade a pursuing predator would benefit from unpredictable patterns of flight behaviour. I believe it likely that many organisms have evolved simple behavioural rules that generate chaotically unpredictable patterns of evading predators.

One thing is certain. Biological systems, from communities and populations to physiological processes, are governed by nonlinear mechanisms. This means that we must expect to see chaos as often as we see cycles or steadiness. The message that I urged more than 10 years ago is even more true today: 'not only in [biological] research, but also in the everyday world of politics and economics, we would all be better off if more people realized that simple nonlinear systems do not necessarily possess simple dynamical properties'.

Further reading

LEON GLASS and MICHAEL MACKEY, *From Clocks to Chaos: The Rhythms of Life*, Princeton University Press, 1988.

ROBERT MAY, 'When two and two does not make four: nonlinear phenomena in ecology', *Proceedings of the Royal Society*, Vol. B228, 1986, p. 241.

8
Is the Solar System stable?

CARL MURRAY

You might be surprised to learn that the Earth's orbit round the Sun, like those of other planets, is chaotic. What does this mean for the future of the Solar System?

People tend to think of the Solar System as a paradigm of order and regularity. We imagine the planets fixed in their orbits around the Sun for all time – an orderly, predictable, unchanging, majestic clockwork that never needs rewinding. We can steer the Voyager 2 spacecraft nearly 5 billion kilometres on a 12-year journey from the Earth to an encounter with Neptune, so it arrives on schedule within kilometres of its target. We can accept unforeseen changes in our everyday lives and even come to terms with natural and man-made disasters, yet we still have faith in the immutability of the orbits of the planets and satellites.

What grounds do we have for such beliefs? Is the behaviour of the Solar System completely predictable, or could the planets ever collide? This is a question that many astonomers have attempted to answer, but it is only in this decade that a better understanding of the problem and a possible solution has emerged. The key to this progress is the study of chaos, whereby even simple, deterministic equations can give complicated unpredictable solutions. Chaos has revealed that our Solar System is not the paragon of predictability that we once imagined.

In the 17th century, Isaac Newton showed that if two bodies attract each other with a force that was proportional to the product of their masses and inversely proportional to the square of the distance between them, then the resulting motion of one body relative to the other would be a precise mathematical

Plate 1 (*above*) Chaos is beautiful. The pattern generated by the chaotic dynamics comes from a formula by I. Gumowski and C. Mira. Points moving around unpredictably produce this highly ordered filigree structure.

Plate 2 Not just a pretty pattern; this beautiful and unusual computer-generated picture shows what happens to equation 1, described by Franco Vivaldi in Chapter 3, when λ is allowed to alternate between two numbers A and B. The colours in the picture show the nature of the motion, ranging from orderly (the dark blue stripes) to completely chaotic (white areas). A is the x-axis and B is the y-axis.

Plate 3 Who was that strange attractor, anyway? The mask-shaped Lorenz attractor (see page 52).

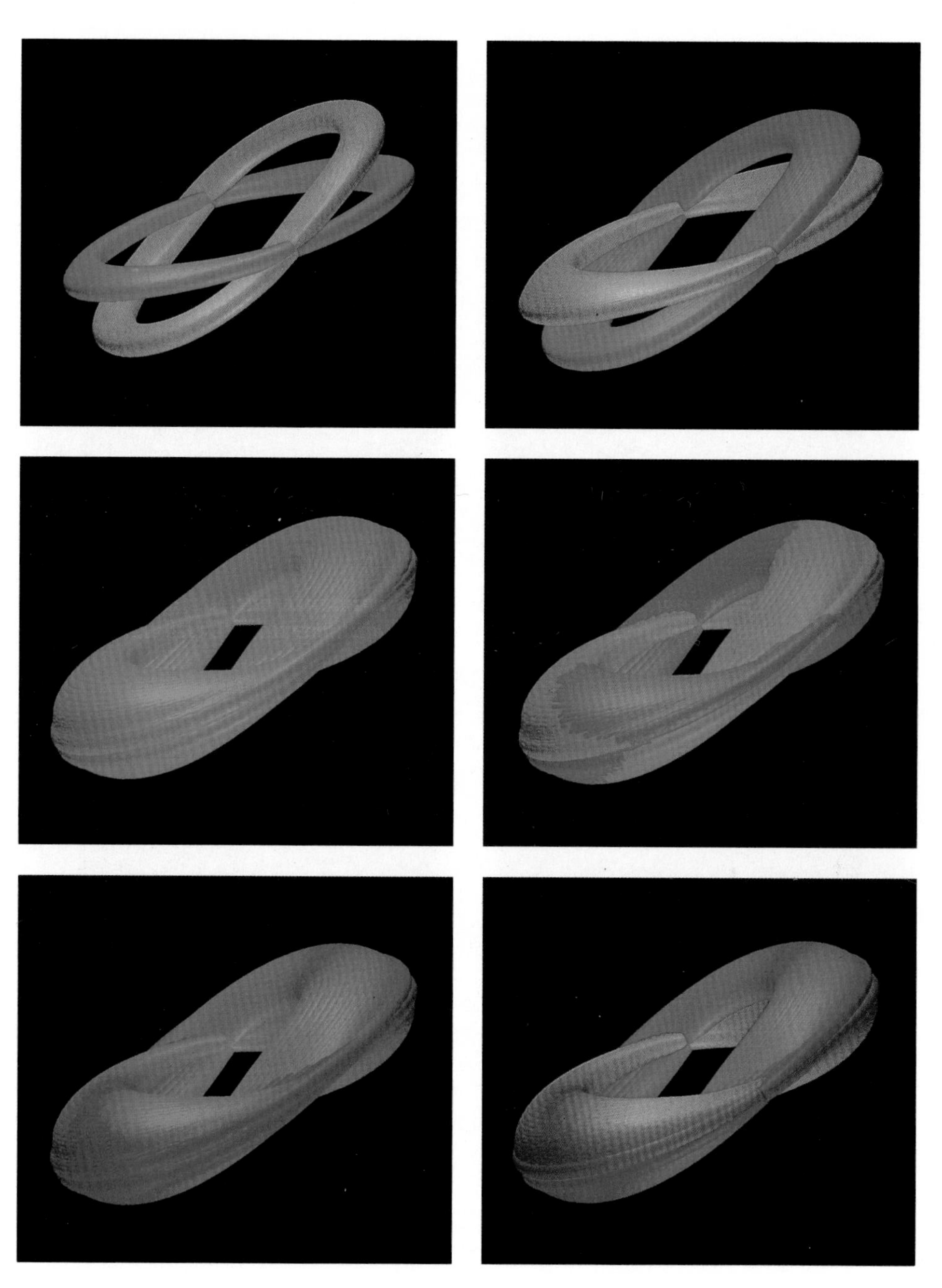

Plates 4 to 9 show the 'phase portraits' of a simple system of coupled pendulums, as described on pages 53–5.

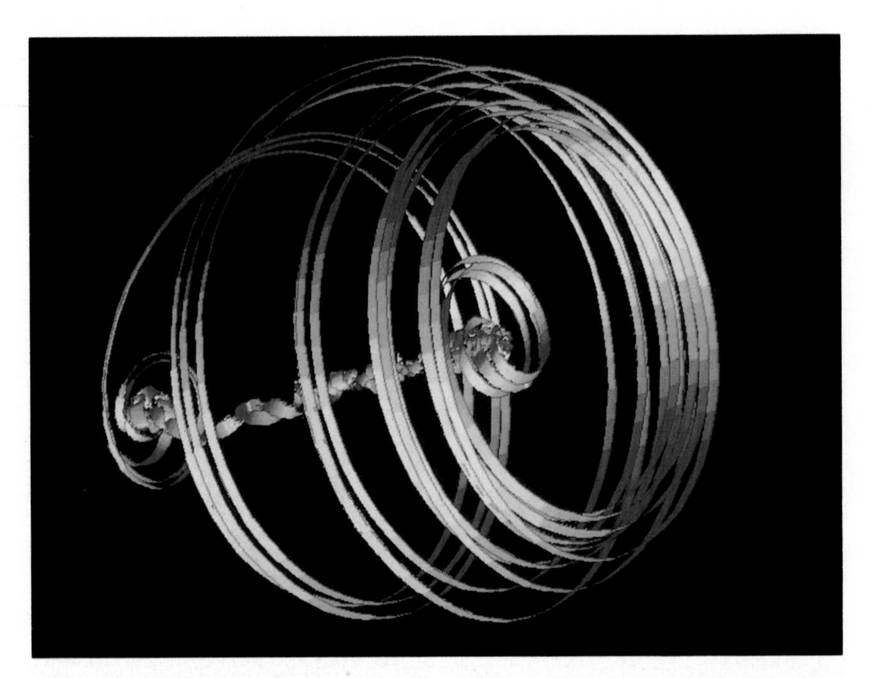

Plate 10 This strange computer graphic shows a reconstructed 'phase portrait' of regular flow in the Taylor-Couette system, as described in Chapter 5.

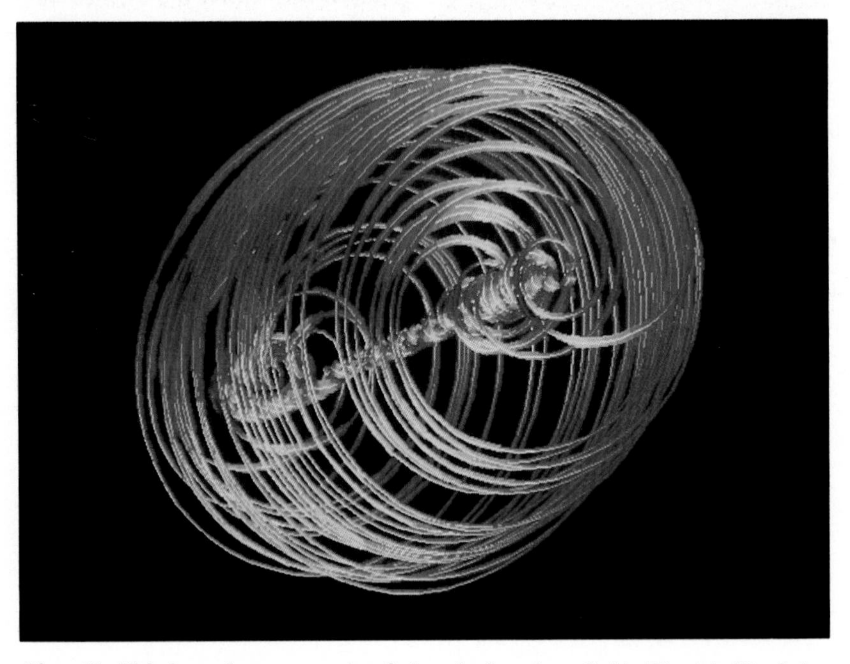

Plate 11 This is a phase portrait of chaotic flow in a fluid. The 'shell' of the structure begins to fill in to give a 'strange attractor'.

Plate 12 (*above*) This is not an oriental emblem but a 'map' of a certain kind of predator-prey system, as described on page 49. The horizontal axis is the prey population and the vertical axis is the predator population, as they vary together over time. The fluctuating populations are mainly found in the golden region, with occasional oscillations into the larger crimson domain.

Plate 13 Pour the ingredients for the spectacular Belousov-Zhabotinskii reaction (as described on pages 111–12) into a Petri dish, and spiral patterns of colour emerge as the reaction undergoes periodic changes. The height of the peaks represents the concentration of the reacting materials.

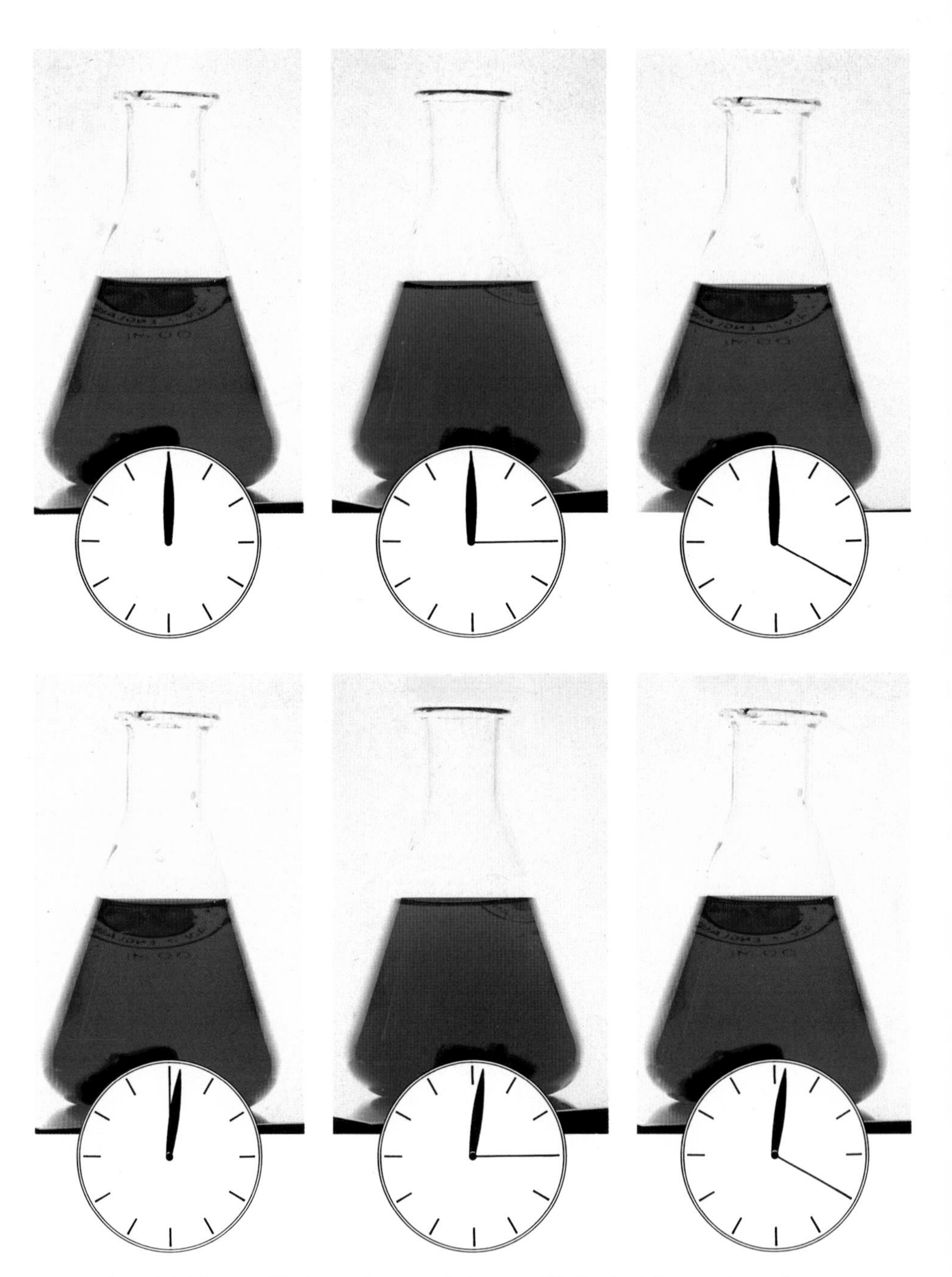

Plate 14 The oscillations in the Belousov-Zhabotinskii reaction show a periodic colour change between red and blue, here shown over a period of about 80 seconds.

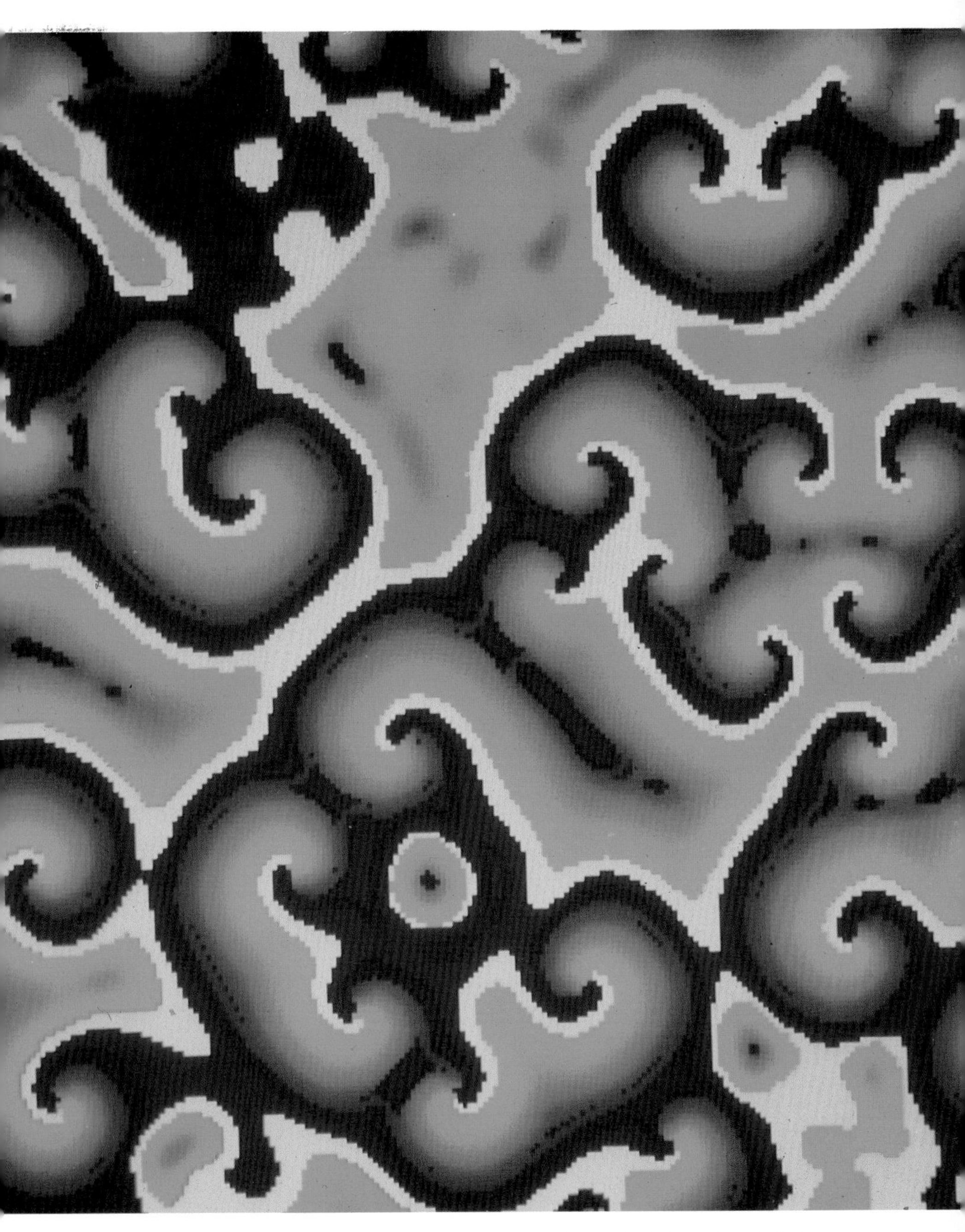

Plate 15 The spiral patterns that emerge from the underlying chaotic behaviour of a metal catalyst (see pages 117–18).

Plate 16 This remarkable picture was created entirely on a computer using fractal geometry (see Chapter 10).

Plate 17 Fractal modelling on a computer produces extraordinary landscapes such as this mountain scene.

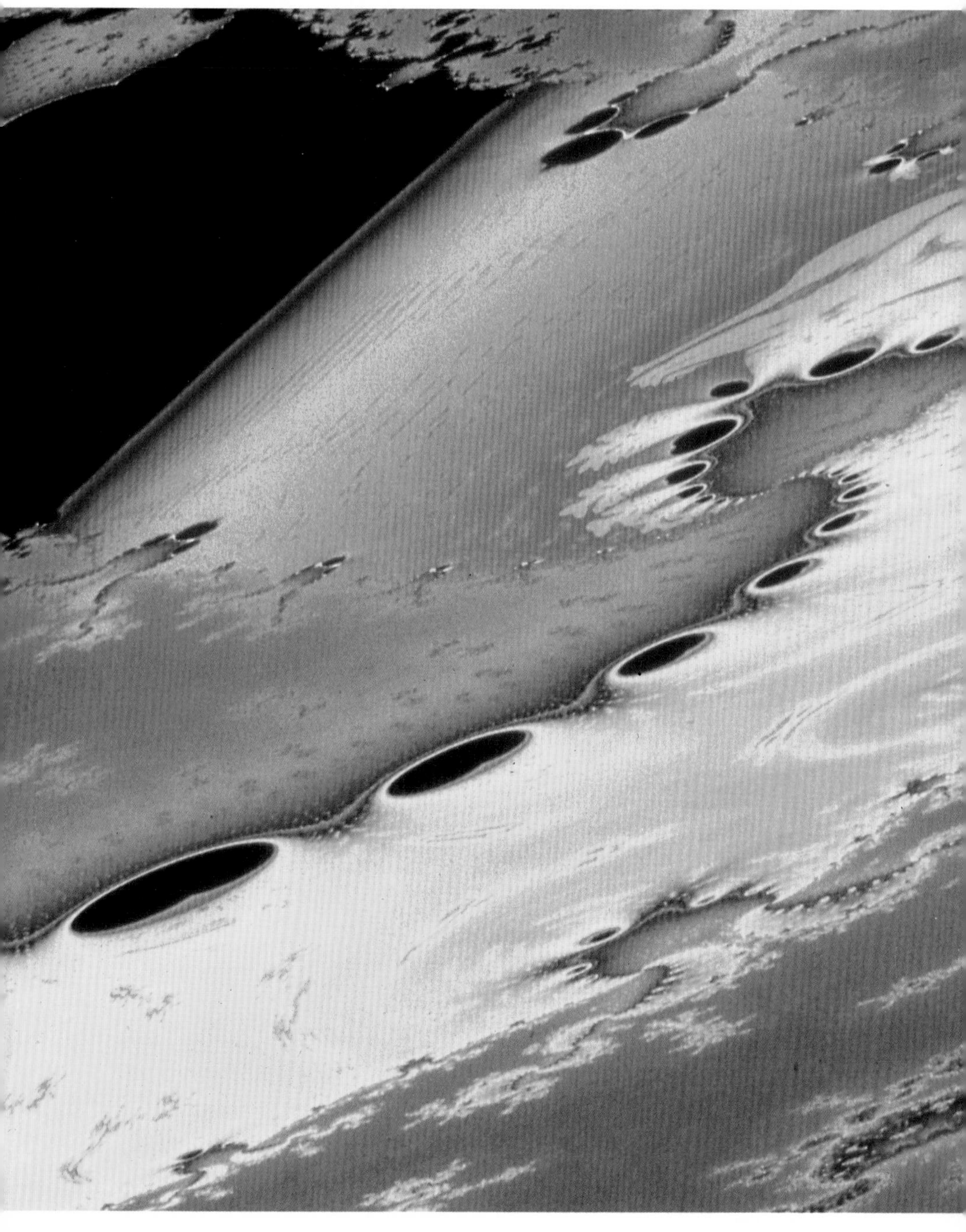

Plate 18 The Mandelbrot set produces rich mathematics and beautiful graphics. Here is a generalized Mandelbrot set, obtained by replacing the usual mathematical formula by a more complicated one (see Chapter 10).

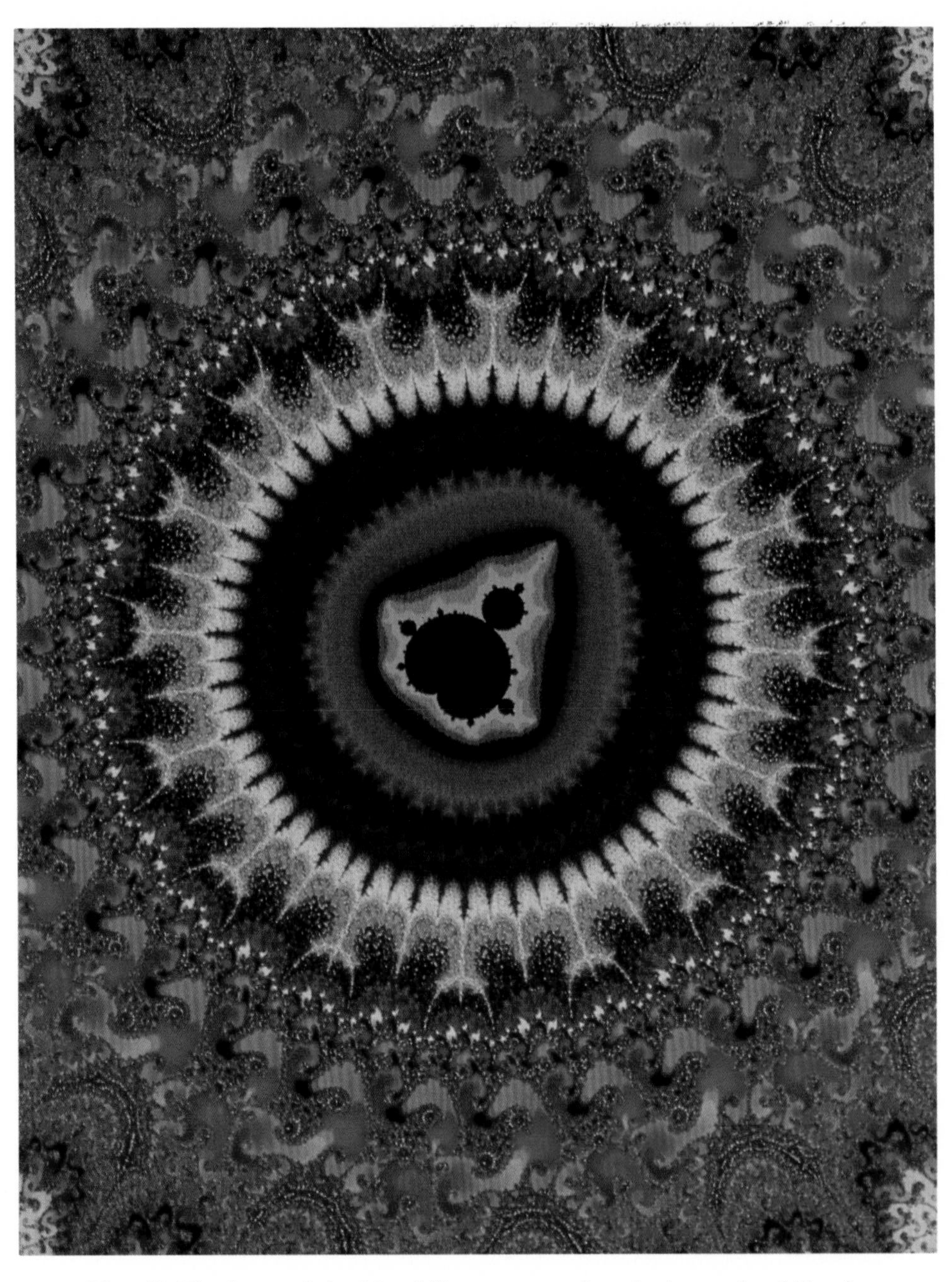

Plate 19 The heart of the Mandelbrot set seen here is the result of blowing up the set by Avogadro's number ($\times 10^{23}$). It remains forever intricate on any scale. In fact, its complication increases as one zooms in.

Plate 20 Not skeins of coloured knitting wool but a relation of the Mandelbrot set, the Julia set, shown here in four dimensions.

Plate 21 (*above*) Fractal reflections on a complex plane produce what are called limit sets (see page 140). Here is a stunning example.

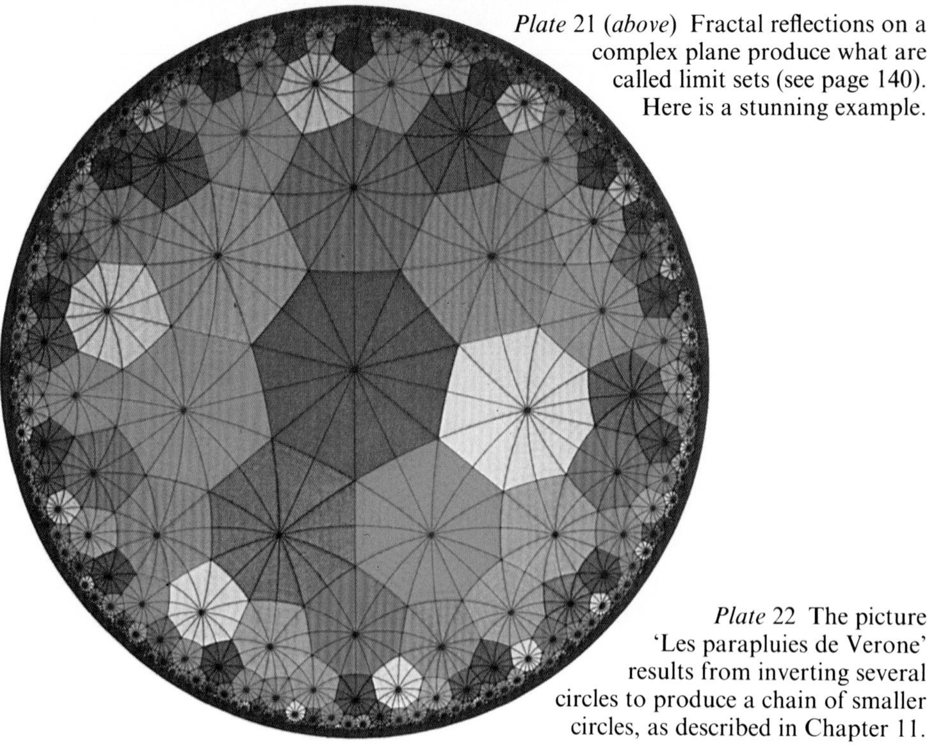

Plate 22 The picture 'Les parapluies de Verone' results from inverting several circles to produce a chain of smaller circles, as described in Chapter 11.

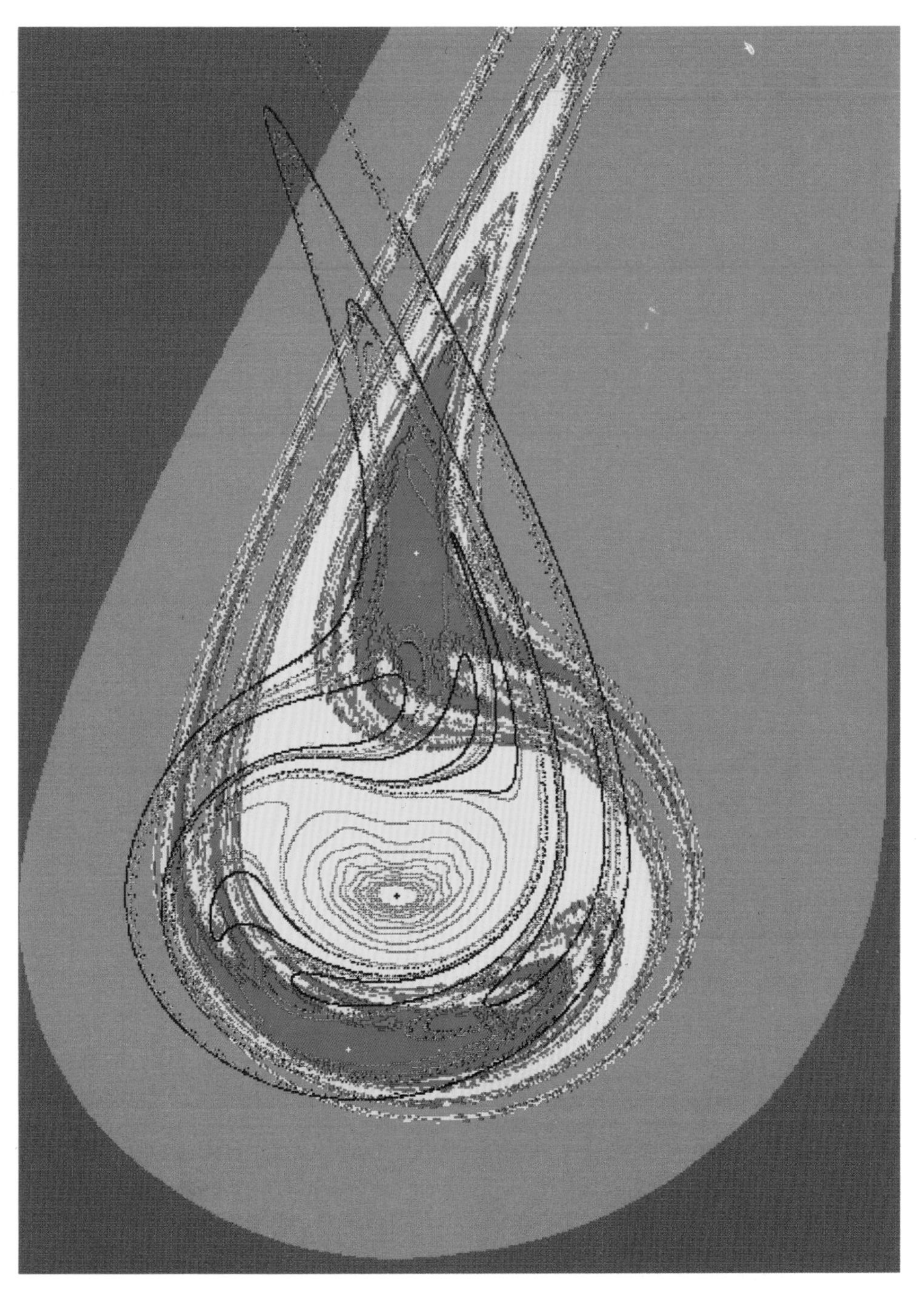

Plate 23 A periodic force causes a nonlinear system to oscillate chaotically. Different colours can be used to map the possible modes of vibration – the attractors – to which the system converges.

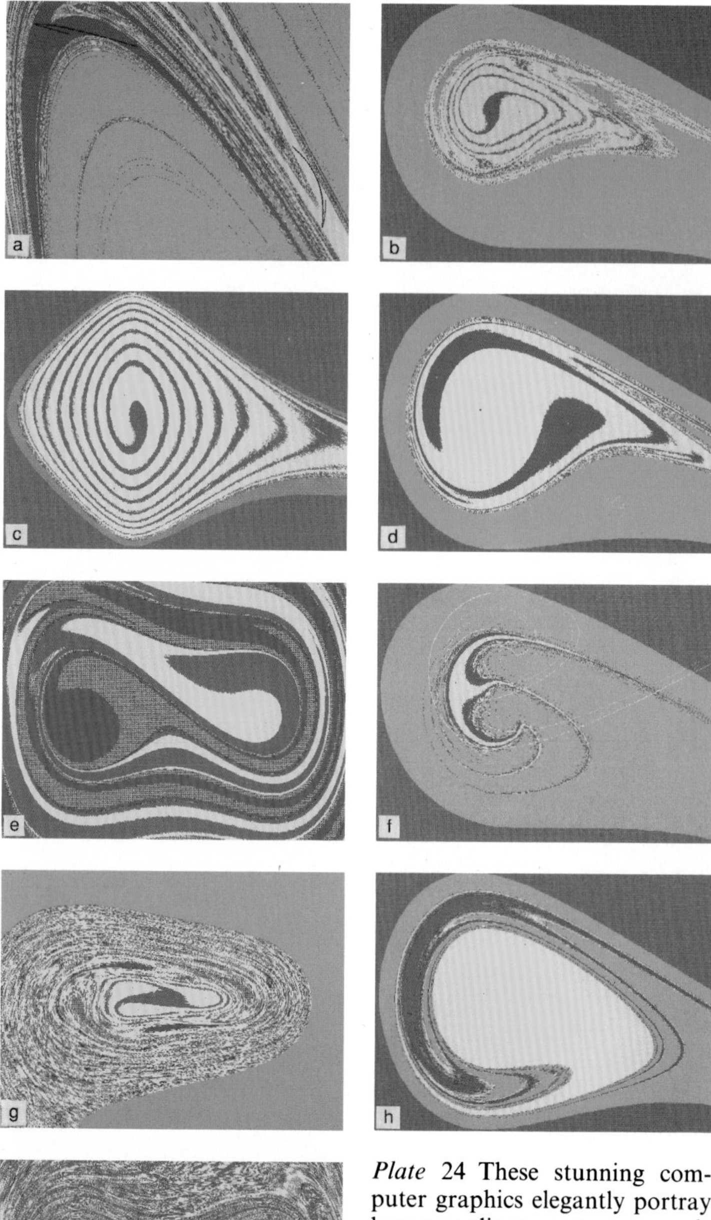

Plate 24 These stunning computer graphics elegantly portray how a nonlinear system responds to a periodic force in a gallery of fractal 'basins'. In most of the graphics, blue signifies disaster, and yellow, safety (see Chapter 12).

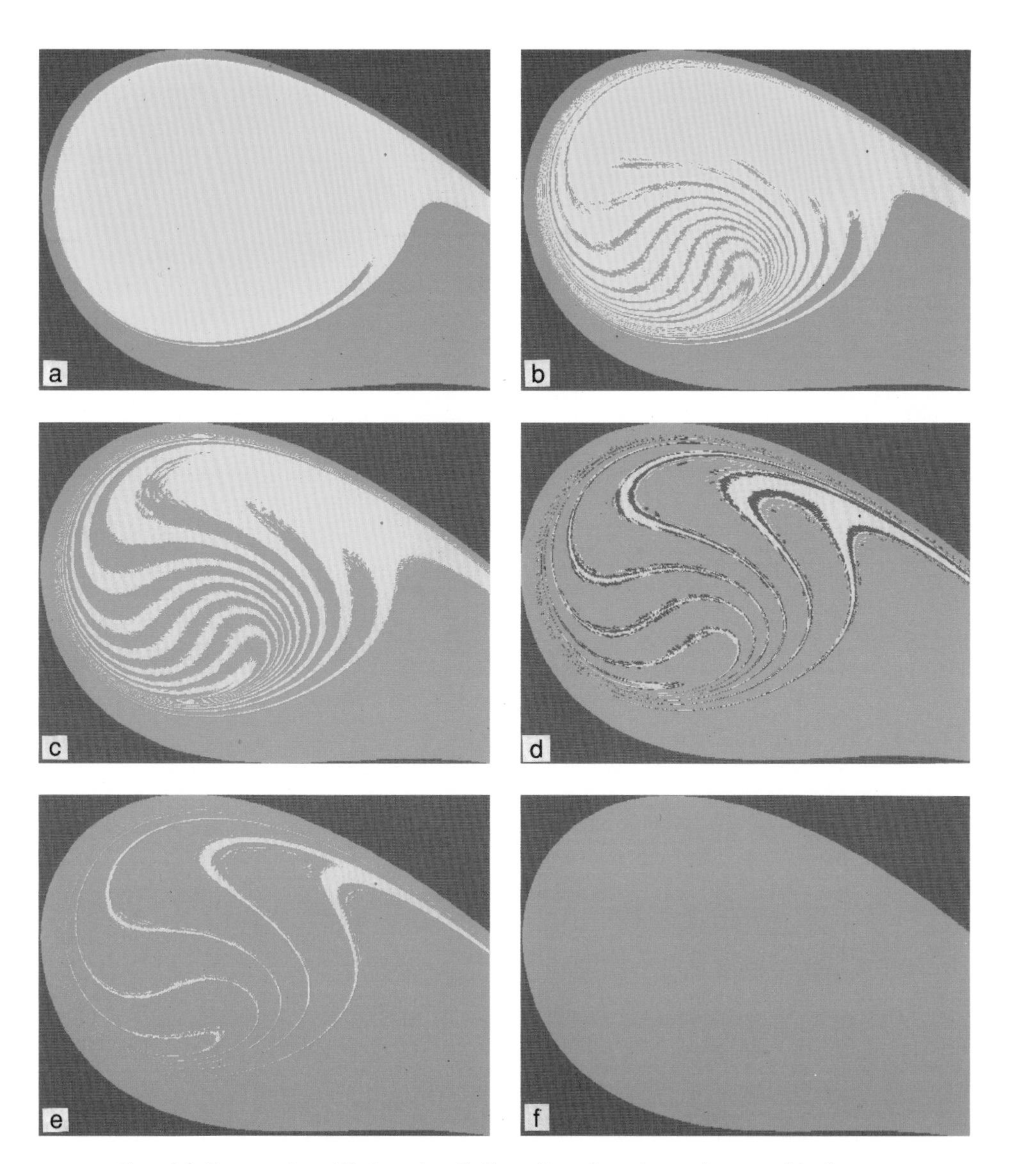

Plate 25 Research at University College London shows how a ship becomes unstable. The fractal pictures above reveal how waves hitting the vessel erode the safe basin of attraction. The first picture shows a large basin which is gradually eaten away by the fractal basin boundary until in the final picture there is no safe basin and no stable solution. This suggests a new 'index of capsizability' which is related to the first intrusion of the fractal fingers into the safe basin (see page 157).

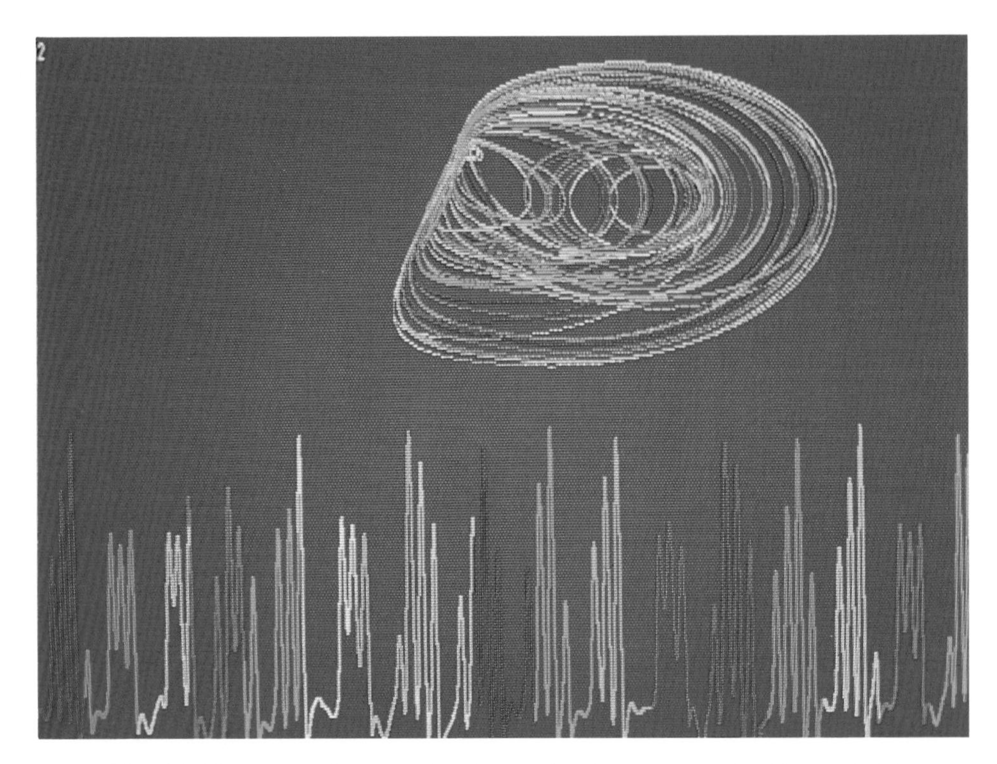

Plate 26 The output from a chaotic circuit when combined with a suitable transmission line of inductors and capacitors. The graph shows how the voltage at the circuit input varies with time if you apply a fixed voltage at the other end of the transmission line. The whirly pattern shows how this voltage varies with the changing current in the last inductor of the line. As you can see, this oscillator does not keep very good time (see Chapter 13).

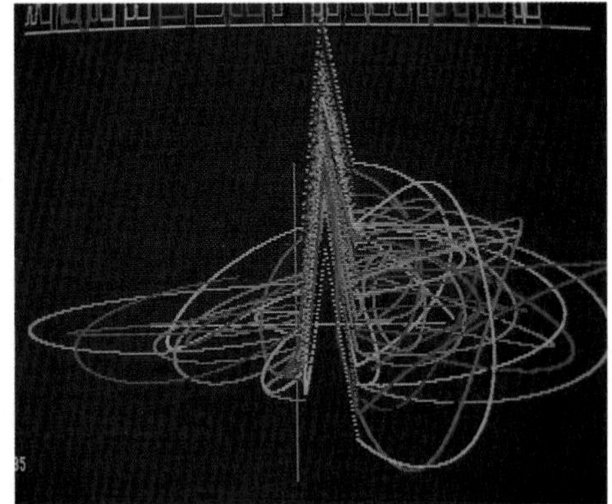

Plate 27 'Phase portraits' of a nonlinear amplifier where two voltages in a feedback arrangement vary together. Note that the pattern never repeats itself exactly. Sometimes the pattern is fairly well behaved (*left*); at other times (*right*) it seems to go crazy. (See Chapter 13.)

curve called a conic section (that is, a circle, ellipse, parabola or hyperbola). Although Newton was not the first person to suggest that the inverse square law of force was responsible for the motion of the planets, his great triumph was to provide a mathematical proof of the consequences of such a law.

He showed that a planet moving under the effects of the Sun's gravity would describe an elliptical path, and that the period of this motion would depend only on the average distance from the Sun. In mathematical terms, he could show that the 'two-body problem' was integrable, in other words, that it was possible to obtain a complete, practical solution to the problem using relatively simple mathematical equations. So, if we are interested only in the two-body problem we can predict any future configuration of the system with arbitrary precision for all time.

But the Solar System is not composed of just two bodies. It is true that the Sun's gravitational field dominates the motion of all the planets and that, to a good first approximation, each planet moves in an elliptical orbit around the Sun. The planets also influence each other's motion, however, all according to the inverse square law. And we can detect these effects although they are small. For example, the basic ellipse of the Earth's orbit is not fixed in space: it gradually rotates, or precesses, at a current rate of 0·3° per century due to perturbations by the other planets, most notably Jupiter.

The French mathematician Pierre Simon de Laplace tried to solve the problem of the Solar System's stability by making some simplifying assumptions about the nature of the gravitational interactions of the planets. Laplace showed that his simplified system was integrable and that there were long-term periodicities (typically, tens of thousands of years) in the movement of the orbits of the planets: he thought he had achieved the elusive analytical solution. Unfortunately, the very terms that Laplace had neglected in his theory were those that could provide possible sources of chaos. So Laplace's proof of stability has to be discounted.

Laplace was one of many scientists who had a fundamental belief that once you had determined the laws governing the

Universe, it was just a matter of solving the equations, with the appropriate starting conditions, to discover its past and future behaviour and that 'nothing would be uncertain'. The study of chaos has revealed that even completely deterministic systems, such as those involving gravitational interactions, can be chaotic and that Laplace's world-view was wrong. For example, the motion of the ball in a spinning roulette wheel is, in principle, a deterministic system. Although the ball and wheel are subject to known forces, trying to predict the final outcome is unlikely to be a rewarding experience.

We now know that, except for special cases, the general motion of many (n) bodies interacting through gravity, the 'n-body problem', is not integrable. A simpler task is to attempt to solve the three-body problem. At the end of the 19th century, the French mathematician Henri Poincaré tackled this problem in some depth. It is clear from his writings that he was aware of the unpredictability of some solutions of the equations of motion. He did not solve the three-body problem; in fact, he proved that a simple, general solution did not exist. However, Poincaré was the first to appreciate the complicated behaviour that could result from the gravitational interaction of just three bodies. He also realized that in the Solar System, chaos and order, stability and instability were closely connected with a phenomenon called 'resonance' (see 'Asteroids and planets in resonance', below).

Resonance pervades the Solar System. It happens when any two periods have a simple numerical ratio. The most fundamental period for an object in the Solar System is its orbital period. This is the time it takes to complete one orbit and depends only on its distance from the central object. For example, the Jovian satellite Io has an orbital period of 1·769 days, which is nearly half that of the next satellite Europa – with a period of 3·551 days. They are said to be in a 2:1 orbit-orbit resonance. This particular resonance has important consequences because the perturbations, resulting from Europa's gravity, force the orbit of Io to become more elongated, or eccentric. As Io moves closer to Jupiter and then further away in the course of a single orbit, it experiences significant tidal stresses resulting in the

Asteroids and planets in resonance

A resonance may happen when there is a simple numerical relationship between two periods that leads to repeated configurations. Consider the case of an asteroid and Jupiter orbiting the Sun. For simplicity, we will take Jupiter's orbit to be circular and assume that all objects orbit in the same plane. An asteroid at the 2:1 Jovian resonance will have an orbital period of six years, which is half Jupiter's period of 12 years. To see how the resonance can be stable or unstable, consider two possible initial configurations with the asteroid and Jupiter aligned with the Sun on the same side of their orbits (conjunction). The asteroid's orbit will be considerably affected by Jupiter if conjunctions occur when the asteroid is at its furthermost point from the Sun (aphelion), because this is where the two orbits are closest.

In the stable case, conjunctions occur at the asteroid's closest point to the Sun (perihelion). The asteroid reaches the danger point of aphelion after three and nine years, but in each case Jupiter is at a different point in its orbit. Every 12 years, the initial configuration repeats itself and so the asteroid always avoids conjunctions at aphelion and is in a stable resonant orbit. In the unstable case, conjunctions always occur at the asteroid's aphelion, leading to a repeated, unstable configuration that cannot continue (see Figure 8.1a).

There is a simple analogy with the motion of a pendulum (see Figure 8.1b). It is stable when you start the bob in the vertically down position. Any small shifts of the bob will cause oscillations

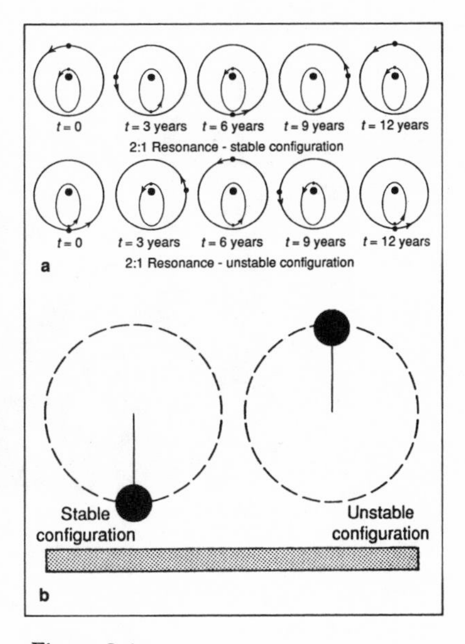

Figure 8.1

about the stable point. However, if you start the bob in the vertically up position, it can remain balanced there only for a short time before it becomes unstable. In fact, the equations of motion of an asteroid in resonance with Jupiter are very similar to those of a simple pendulum. To introduce chaos into the pendulum system we need only to oscillate the point of suspension in a regular fashion. The equivalent action in the Sun-Jupiter-asteroid system would be to make the eccentricity of Jupiter's orbit nonzero.

---

active volcanoes that Voyager observed. To make things more complicated, Europa is simultaneously in a 2:1 resonance with the next satellite Ganymede, the trio being involved in an intricate configuration which Laplace had studied.

In the Saturnian system, resonant pairs of satellites include Mimas and Tethys, Enceladus and Dione, and Titan and Hyperion. Resonances are curiously absent from the Uranian

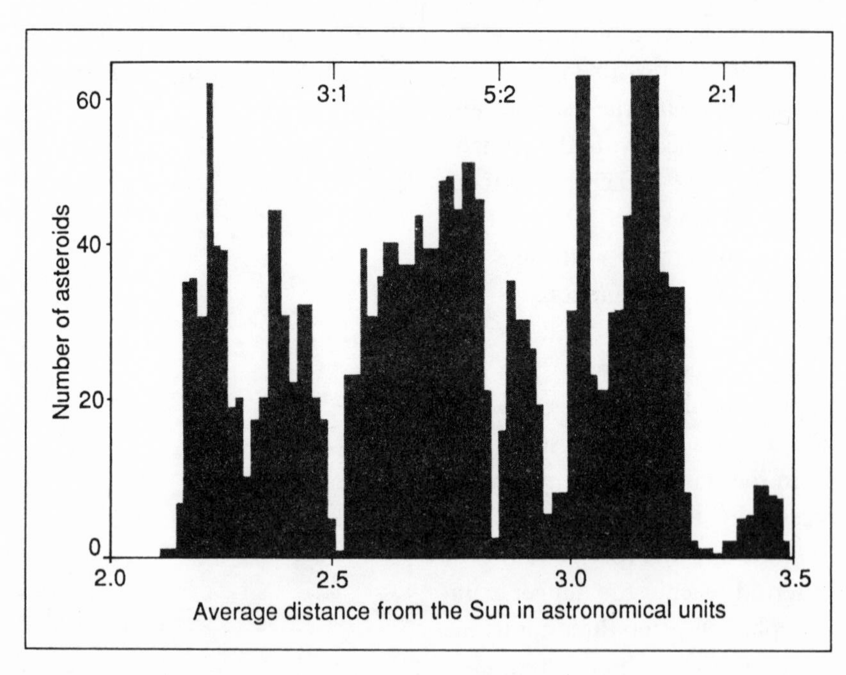

Figure 8.2 Gaps in the asteroid belt occur at locations corresponding to resonances with Jupiter.

satellite system, although there have been recent explanations that invoke the effects of chaos. Among the planets, Jupiter, with a period of 11·86 years, and Saturn, with a period of 29·46 years, are close to a 5:2 resonance, yet the only true orbit-orbit resonance between planets is a curious 3:2 resonance between Neptune and Pluto. C. J. Cohen and E. C. Hubbard at the U S Naval Weapons Laboratory discovered this in 1965, not by observation, but by a 'numerical integration' of the system, When the equations of motion of a system cannot be solved analytically by mathematics, a possible alternative is to solve them numerically using a digital computer. Although such a numerical integration provides less insight than a mathematical solution, it is one of the most powerful tools in modern dynamical astronomy.

The most likely reason that resonance is so common in satellite systems is due to the effects of tides. As a satellite raises a tide on a planet, there is an exchange of angular momentum between the bodies, resulting in a change in the orbit of the satellite and in the spin of the planet. Consequently, the orbits of the natural satellites today may bear little resemblance to their original ones; this is certainly true in the case of the Moon. As a satellite evolves, its orbital period changes and it may encounter a resonance with another satellite. In certain circumstances, the satellites become locked in a resonance and continue to evolve tidally, so maintaining the resonant configuration. The planets do raise tides on the Sun but these are not so important because of the greater relative distances involved. This may explain the lack of orbit-orbit resonances in the planetary system.

The most striking example of resonance occurs in the asteroid belt, a collection of more than 4000 catalogued objects orbiting between Mars and Jupiter. In 1867, the American astronomer Daniel Kirkwood noticed that the asteroid orbits were not randomly distributed. There were distinct gaps in the belt at locations that corresponded to resonances with Jupiter. For example, there were no asteroids at the 3:1 resonance – a distance of 2·5 astronomical units – or at the 2:1 resonance at 3·3 astronomical units (see the diagram opposite). This was the opposite situation to that found in the satellite systems where there was a preference for objects in resonance. Although there were fewer than 100 asteroids known in Kirkwood's day, his conclusions concerning the association with Jovian resonances were correct, but astronomers have only recently explained the gaps satisfactorily.

Astronomers had proposed several theories, ranging from the suggestion that the gaps were a kind of dynamical 'illusion', to the idea that increased collisions at the resonances had caused asteroid material to wear down, producing objects too small to be observed. In 1981, Stan Dermott and I began studies at Cornell University on how the asteroid orbits were distributed to try to find out which theory was correct. We concluded that you could explain the gaps in terms of a simple three-body problem involving the Sun, Jupiter and an asteroid. But we still lacked a mechanism that could actually remove an asteroid from a resonance, although we recognized that we could tackle the problem on a computer, by solving the equations of motion to study the behaviour of an asteroid over millions of years. The main drawback of such an approach was the expense – computer time is not cheap.

The breakthrough came in 1981 when Jack Wisdom, then a PhD student at the California Institute of Technology, developed a new numerical method for studying the motion of asteroids at resonance. Wisdom knew about chaotic dynamics, and in particular how to derive 'mappings' to speed up the numerical work. Given the state of a system at a certain time, a mapping could give a precise, algebraic method for calculating the state at some fixed time interval later. He still had to carry out the mapping on a computer, but using the mapping speeded up calculations 1000 times. As part of his thesis work, Wisdom derived a mapping to study motion at the 3:1 Jovian resonance. He showed that asteroids, moving under the gravitational effects of the Sun and Jupiter, at this resonance could undergo large, unpredictable changes in their orbits. Such orbits were chaotic. Wisdom went on to show that these orbits crossed the orbit of Mars and would eventually impact or be scattered by the planets.

This was the mechanism of removal that astronomers had been seeking.

Wisdom showed that there was an extensive chaotic zone at the 3:1 resonance, which matched the observed width of the Kirkwood gap. This discovery was to solve another problem in Solar System dynamics. Most people think that meteorites are fragments of asteroids that eventually collide with the Earth. You can measure the 'age' of meteorites by finding out how long they have been exposed to the cosmic rays in the Solar System. Members of one particular class of meteorite, the chondrites, have very short exposure ages of only a few million years. George Wetherill, of the Carnegie Institution in Washington DC, had shown that these meteorites had to come from the vicinity of the 3:1 resonance, but he lacked a mechanism. Wisdom provided the mechanism and carried out numerical integrations to show that the chaotic orbits of objects at the 3:1 resonances could become eccentric enough for them to start crossing the Earth's orbit.

The whole problem of where objects in Earth-crossing orbits come from is more than an abstract academic question. The impact of a large asteroid on the Earth would be one of the worst natural disasters that our planet could face. The 3:1 gap is not entirely clear of asteroids. At least two asteroids in the gap, Alinda and Quetzalcoatl, are actually in resonance with Jupiter, and 1989 AC, the asteroid that will pass within just 0·011 astronomical units of the Earth in 2004, is also probably in resonance with Jupiter at the gap. It is important for dynamicists studying the Solar System to understand where such objects come from and how they evolve. This requires a knowledge of chaos.

The two periods involved in a resonance relation do not have to be orbital periods. Another common form of resonance in the Solar System is spin-orbit resonance, where the period of spin of an object (the time it takes the orbit to rotate once about its own spin axis) has a simple numerical relationship with its orbital period. For example, Mercury is locked in a 3:2 spin-orbit resonance. A more obvious example is our own Moon, which is in synchronous rotation because of the 1:1 spin-orbit resonance that forces it to keep the same face towards the Earth. The far side of the Moon was completely hidden from us until the era of spaceflight. Most natural satellites in the Solar System are in synchronous spin states although this was not their original state: they have evolved into such configurations because of tidal effects. A simple theory allows us to predict the timescales for evolution into the synchronous state. The timescale depends on the mass of the satellite and its distance from the central object.

Prior to the Voyager encounters with Saturn, people had wondered whether or not the satellite Hyperion was in synchronous rotation. After all, it is small and one of the most distant of the Saturnian satellites. Observations from Voyager 2 revealed an irregularly shaped object shaped rather like a hamburger, or a potato. Measurements of Hyperion's rotation made by Voyagers 1 and 2 suggest a spin period of 13 days, compared with an orbital period of 21 days, so Hyperion does not appear to be in an obvious spin-orbit resonance.

In 1984, Jack Wisdom and Stan Peale working at the University of California, Santa Barbara, and François Mignard of CERGA, Grasse, published a classic paper in which they showed that the simple theory worked out for satellite rotations does not apply to Hyperion, because it is distinctly nonspherical. Hyperion's rotation is certainly not synchronous, but neither is it regular; it is chaotic. Furthermore, Wisdom, Peale and Mignard showed that Hyperion is also 'attitude unstable', which means that its spin axis is not fixed and the satellite is tumbling in space as well as rotating chaotically. In normal circumstances, the satellite orbit would become more circular and eventually the chaotic behaviour would disappear, but, ironically, tiny Hyperion is locked in an apparently stable 4:3 orbit-orbit resonance with the massive satellite of Saturn, Titan. This forces Hyperion's orbit to be eccentric rather than circular, so the chaos persists, resulting in a satellite with a chaotic spin but a regular orbit.

So chaotic motion does exist in the Solar System in a variety of forms. But are the orbits of the planets chaotic? The answer to this question is likely to come from long-term integrations of the planetary system using a new generation of digital computers. In this decade, there have been a number of separate efforts to investigate the motions of Jupiter, Saturn, Uranus, Neptune and Pluto. The inner planets are notoriously difficult to include in such integrations because very small time-steps are needed to follow them accurately.

An international consortium of Solar System dynamicists led by Archie Roy of the University of Glasgow, carried out Project LONGSTOP (Long-term Gravitational Study of the Outer Planets) on the Cray supercomputer at the University of London. This involved integrating the orbits of the outer planets for 100 million years. Its results revealed several curious exchanges of energy between the outer planets, but no signs of gross instability.

Another project involved constructing the Digital Orrery by Gerry Sussman and his group from the Massachusetts Institute

Figure 8.3 The Digital Orrery numerically integrates orbits in the Solar System. It is attached to a computer workstation at the Massachusetts Institute of Technology and can study the motions of 10 gravitationally interacting bodies at 60 times the speed of a VAX minicomputer.

of Technology. The group used this machine, which has a computer architecture designed to mimic the interactions between the planets, to integrate the orbits of the outer planets over 845 million years (some 20 per cent of the age of the Solar System). In 1988, Sussman and Wisdom produced integrations using the Orrery which revealed that Pluto's orbit shows the tell-tale signs of chaos, due in part to its pecular resonance with Neptune. This does not mean, however, that the resonance is unstable or that Pluto and Neptune could ever collide, even though their orbits intersect. Recent work suggests that this chaos arises from resonances within resonances; these can limit the extent of Pluto's wandering and preserve the main resonance with Neptune.

If Pluto's orbit is chaotic, then technically the whole Solar System is chaotic, because each planet, even one as small as Pluto, affects the others to some extent through gravitational interactions. But we now realize that although chaos means that some orbits are unpredictable, it does not necessarily mean that planets will collide – chaotic motion can still be bounded. In 1989, Jacques Laskar of the Bureau des Longitudes in Paris published the results of his numerical integration of the Solar System over 200 million years. These were not the full equations of motion, but rather averaged equations along the lines of those used by Laplace. Unlike Laplace, however, Laskar's equations had some 150 000 terms. Laskar's work showed that the Earth's orbit (as well as the orbits of all the inner planets) is chaotic and that an error as small as 15 metres in measuring the position of the Earth today would make it impossible to predict where the Earth would be in its orbit in just over 100 million years' time.

Laskar's results still have to be confirmed by integrating the full equations of motion, but this will have to wait until the next generation of supercomputers arrives. Meanwhile, we can take comfort from the fact that his work does not imply that any orbital catastrophe awaits our planet, only that its future path is unpredictable. It seems likely that the Solar System is chaotic but nevertheless confined, although we have yet to prove it. More than 300 years after the publication of Newton's *Principia*, we are still struggling to understand the full implications of his simple inverse square law of gravity. We have begun to view our Solar System of chaos in a light that is revealing the true intricacies of its majestic clockwork.

Further reading

ANITA M. KILLAIN, 'Playing dice with the Solar System', *Sky and Telescope*, Vol. 72, 1987, p. 241.

9

Clocks and chaos in chemistry

STEPHEN SCOTT

Most chemists believe that the course of a chemical reaction is always predictable. But some catalytic reactions in both inorganic and organic chemistry can behave in bizarre and unruly ways.

Chemistry provides some of the most well-defined examples of chaos. The dynamics of chemical reactions can show the same kind of periodic and chaotic behaviour that appears in population dynamics in biology. But the transition from order to chaos is easier to study because we can readily control the conditions of the experiments. The results are often visually dramatic, with sudden colour changes – of course, the timescales are much shorter.

The equations that describe how fast a chemical reaction proceeds are nonlinear. For a chemical reaction to happen at all, molecules have to approach each other, so the more molecules there are the faster the reaction goes. The rate of reaction, therefore, depends on the concentrations of the reacting molecules, normally raised to some power (the square of the concentration, for example, if two molecules are involved in the reaction). Sometimes, the reaction rate also depends on the concentrations of the molecules formed from the reaction. These molecules may speed up the reaction, so acting as autocatalysts, resulting in positive feedback.

Combustion, or oxidation, provides a good example of positive feedback. In this case, it is the heat produced by the reaction that ever-increasingly speeds it up. Everyone knows that above a certain temperature objects can suddenly burst into flame, burn furiously, and then, when the heat produced drops off, the flames can suddenly cease. Sometimes, the reaction can choose between two stable modes. For instance, carbon monoxide will burn in oxygen, in the presence of a platinum catalyst, to give carbon dioxide. (This is the way catalytic convertors in car exhausts work.) The reaction varies with the catalyst's temperature to give two branches: one corresponding to high reaction rates, the other to low rates. Over some range of temperature, these branches overlap, leading to two possible rates for identical experimental conditions. The reaction can also oscillate between the two branches. In similar reactions, chaos can ensue – as we shall see later.

Another kind of chemistry where the rate can increase or decrease dramatically and sometimes discontinuously in response to small smooth changes in experimental conditions are the intriguing 'clock' reactions. An example is the Landolt reaction, in which dissolved iodate (IO_3^-) and iodide (I^-) ions react together to form iodine (I_2). This product does not survive long, however, because it oxidizes the other main ingredient of the reaction, bisulphite ion (HSO_3^{2-}), to form sulphate (SO_4^{2-}) and is itself reduced back to (more) iodide. This provides positive feedback; the rate of the first reaction increases as the concentration of iodide increases. As the reaction proceeds further the bisulphite keeps the concentration of iodine low, until the bisulphite is almost all consumed. The moment that the bisulphite is exhausted there is a rapid build up of iodine. If we add starch, as an indicator for iodine (starch and iodine form a deep blue compound), we can see what happens. Initially, the solution is colourless, and stays so for a while, then suddenly goes blue as the iodine forms. If bisulphite is in excess, the high concentration of iodine is only transient, and the solution 'blinks' blue and then turns colourless again.

Positive feedback can also cause chemical reactions to oscillate spectacularly. The famous Belousov-Zhabotinskii, or B-Z, reaction not only shows oscillations but also displays chaos under certain conditions. Although chemists have studied its intricate dynamics for many years, they still do not fully understand the underlying mechanism.

Thirty years ago, Boris Belousov, who was at the USSR Ministry of Health in Moscow, noticed quite by chance that mixing a cocktail of chemicals – sulphuric acid, potassium bromate, cerium sulphate and malonic acid – produced oscillations in the concentrations of bromide (initially present as an impurity but also produced in the reaction) and cerium ions. But poor Belousov had trouble convincing other chemists that the periodic behaviour was a genuine phenomenon; they blamed inefficient mixing of the reactants. The problem was that chemists assumed that such behaviour – where a reaction couldn't make up its mind which way to go, refusing to settle down to an energetically stable state – contravened the second law of thermodynamics.

When Belousov submitted his work to an academic journal for publication, he was told that his 'supposedly discovered discovery was impossible' and that he would have to do a lot more work before the journal could accept the paper. Six years later, after much extra work and revision, the paper was rejected as 'too long'. Thoroughly disheartened, Belousov vowed never to publish, although he was persuaded to produce a short report which appeared in an obscure conference proceedings on radiation medicine.

Some years later, Anatol Zhabotinskii at Moscow State University learnt of the curious reaction and, through careful research, was able to show that the reaction was genuine. The work was then published. Now there are many volumes on such clock reactions. Belousov, alas, did not live to see his reaction accepted, nor to collect the 1980 Lenin prize awarded to him, with Zhabotinskii and three others.

Basically, the B-Z reaction is a more complicated version of the Landolt reaction. It involves oxidizing an organic compound instead of bisulphite, this time with potassium bromate and bromide in the presence of a catalyst (see 'The amazing Belousov-Zhabotinskii reaction', below, for a recipe and details of the underlying chemistry). The catalyst is usually a mixture of cerium ions in two oxidation states, cerium(III) and cerium(IV), and an iron compound called ferroin. This acts as a visual indicator for

The amazing Belousov-Zhabotinskii reaction

Here is a tried and tested recipe for this oscillatory reaction:

500 millilitres of sulphuric acid (1 molar)
14·30 grams malonic acid
5·22 grams potassium bromate
0·548 grams ammonium ceric nitrate
1–2 millilitres of ferroin (0·025 molar)

(Make the ferroin by dissolving 1·485 grams of 1,10-phenanthroline and 0·685 grams of hydrated ferrous sulphate in 100 millilitres of water.)

Stir the mixture continuously. The resulting oscillations between blue and red with a period of about 1 minute will last for several hours.

If you pour some of the mixture as a thin layer into a Petri dish, beautiful patterns, with blue concentric circles, called targets, on a red background will develop.

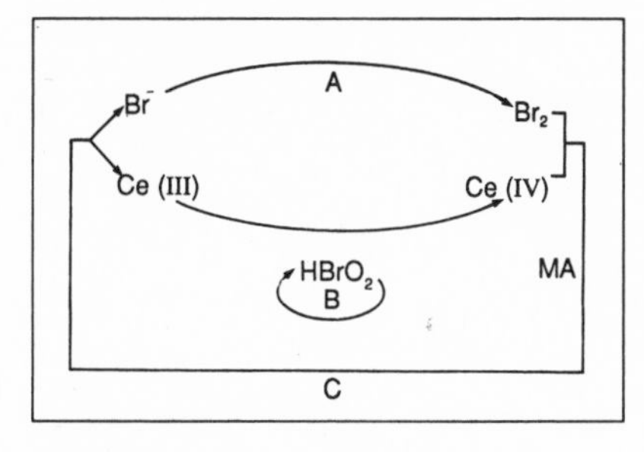

Figure 9.1 Chemical cycles in the B-Z reaction.

We can split the reaction into three overall processes (see Figure 9.1). In process A, bromate ions oxidize bromide ions to produce bromine.

1. $BrO_3^- + Br^- + 2H^+ \rightarrow HBrO_2 + HOBr$
2. $HBrO_2 + Br^- + H^+ \rightarrow 2HOBr$
3. $HOBr + Br^- + H^+ \rightarrow Br_2 + H_2O$

As the concentration of bromide ions decreases, so does the rate of step 2. The bromate ions then compete for reaction with the hypobromous acid ($HBrO_2$) and switch the system to process B. In this part of the reaction, cerium oxidizes from oxidation state (III) to oxidation state (IV). This gives the colour change from red to blue.

4. $BrO_3^- + HBrO_2 + H^+ \rightarrow 2BrO_2 + H_2O$
5. $BrO_2 + Ce(III) + H^+ \rightarrow HBrO_2 + Ce(IV)$
6. $2HBrO_2 \rightarrow BrO_3 + HOBr + H^+$

Because two BrO_2 radicals are produced in step 4, and each reacts rapidly to form an $HBrO_2$ molecule, this part of the reaction constitutes an autocatalytic cycle. The autocatalysis causes the rate of this process to increase very quickly once it has switched on, so red changes rapidly to blue. The growth in the concentration of $HBrO_2$ is limited by step 6. The switch between processes A and B will occur when the rates of steps 2 and 4 are approximately equal.

The final stage, process C, must regenerate the bromide ion and reduce the catalyst back to its lower oxidation state. We do not understand this part of the reaction but we can use the following representation. Malonic acid (MA) reacts with bromine to give bromomalonic acid (BrMA). If this is then oxidized by the cerium (IV), we will regain bromide and cerium (III). The oxidized form of the catalyst can also react directly with malonic acid, so we may get fewer than one bromide ion per cerium (III) ion produced.

So we describe process C as:

7. $MA + Br_2 \rightarrow BrMA + Br^- + H^+$
8. $2Ce(IV) + MA + BrMA \rightarrow$ $fBR^- + 2Ce(III)$ and other products,

where f is known as the 'stoichiometric factor'. In the simplest computer analyses, f is assumed to be a constant, say $f \approx 1$.

In modelling used to match complex oscillations, we attempt to allow f (the number of bromide ions produced as two cerium ions are reduced) to be a function of the instantaneous concentrations of other species, such as HOBr. Process C sees the blue change to red and resets the chemical clock for the next oscillation.

---

the reaction, being magenta when the iron is in its reduced state (Fe^{2+}) and blue when it is oxidized (Fe^{3+}). Plate 14 shows a typical B-Z mixture at different times during a period of about 80 seconds, oscillating in composition and thus colour.

The simplest way of carrying out the B-Z reaction is to operate under 'batch' conditions. Here we just mix the reacting materials in a beaker at the start of the reaction and watch what happens. This produces long trains of perhaps 100 oscillations. This approach has important disadvantages. Eventually, the reactants get used up, so the reaction is proceeding against a continuously, if only slowly, changing background. Each oscillation will, therefore, be slightly different from its predecessor. Eventually, the oscillations will stop completely as the reaction approaches its end.

If we want to study the dynamics in any detail, we obviously want to keep the reaction going as long as possible by maintaining it far from equilibrium. To do this, we drive the reaction by continually pumping in fresh reactants, into what is called a continuous flow reactor (with a corresponding outflow of products to maintain a constant volume). Flow reactors are ideal for studying autocatalytic reactions. In a series of experiments, we might vary the concentration of each of the reactants flowing in, or the rate at which the solutions are being pumped through the reactor (the flow rate).

Under these conditions, the B-Z reaction can show immense complexity, and we can study the potential routes, through periodic oscillations to chaotic dynamics. At low rates of flow, the oscillations resemble those shown in Plate 14 from a batch reactor – simple, large-amplitude, low-frequency waves. At high flow rates, the waves have a much smaller amplitude and higher frequency. If we increase the pumping rate sufficiently, the oscillations may become damped out, so obtaining a steady reaction. It is between these extremes of flow that we find the most interesting behaviour. We can, for example, obtain the kind of period doubling found in fluid flows and biological populations, as other chapters have described (see Figure 9.4). But more significant is the route to chaos through 'mixed-mode' oscillations. The steps up in complexity can be neatly represented by a so-called 'Devil's staircase' – a fractal object with chaos in between its treads (see 'Climbing the Devil's staircase', below). Mixed-mode oscillations consist of large and small peaks, and the Devil's staircase plots how the fraction of small peaks varies with the experimental conditions.

The evidence that chaotic patterns appear in such experiments now seems incontrovertible. Researchers have found 'strange attractors' and other phenomena typical of chaotic dynamics described elsewhere in this book. There is still an active debate as to where this chaos really comes from. The question some chemists are asking is whether the chaos arises solely from the

Climbing the Devil's chemical staircase

In the B-Z reaction, the oscillations in colour changes can be extremely complicated. The transition in waveform from simple, large to simple, small amplitudes is by no means straightforward. An almost unlimited number of 'mixed-mode' oscillatory patterns exists.

These complex waveforms have differing numbers of large and small waves in each complete period. Figure 9.2 shows some examples. To help to catalogue the various periodicities, we can label each waveform as an L^S pattern, where L is the number of large excursions and S the number of small peaks in a full repeating unit. Thus the patterns in Figure 9.2 are 2^3, 2^8 and 1^4. The first two could also be denoted $1^1 1^2$ and $1^4 1^4$ respectively.

Another quantity, related to this notation, is called the firing number, N. We can define this as the ratio of the number of small peaks to the total number of peaks per cycle: $N = S/(L+S)$. The simple, large amplitude oscillation at low flow-rate, the 1^0 form, has $N=0$; at the high flow-rate end, the small amplitude peaks, 0^1, have $N=1$.

If we start with the 1^0 pattern and slowly increase the flow rate, the first of the mixed-mode oscillations encountered is usually that with one large peak and one small peak: a 1^1 with $N=0{\cdot}5$. This form will exist for a finite range of flow rate, and will be succeeded by other patterns. We may discover a sequence in which the repeating

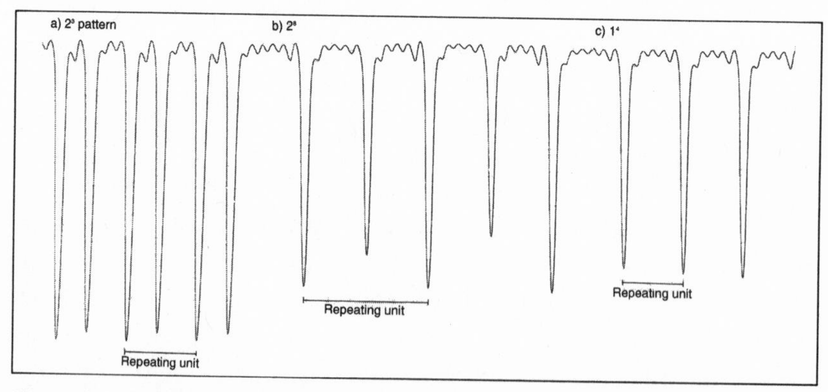

Figure 9.2 Complex oscillatory patterns for the B-Z reaction.

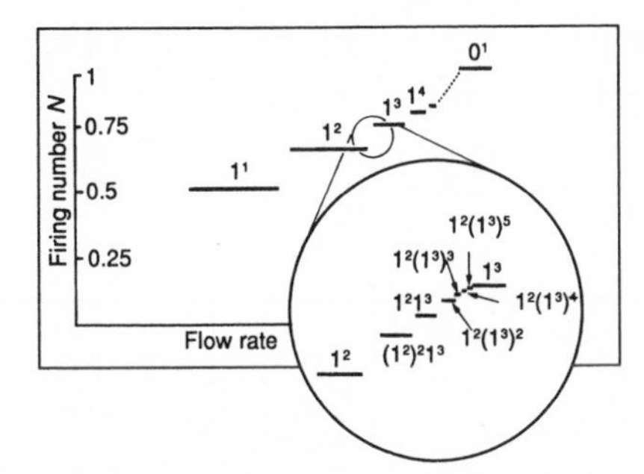

Figure 9.3 A Devil's staircase of firing number and flow rate.

unit gains one extra small peak at each stage, so we see 1^2, 1^3, 1^4 and so on – with N jumping through 0·66, 0·75, 0·8 respectively.

If the firing number is plotted against the flow rate, as shown in Figure 9.3, the result is a series of steps which form a 'Devil's staircase'. The idea of adding one extra small peak to produce a new pattern is given a firmer mathematical basis through 'Farey arithmetic'. A Farey sum of two rational numbers P/Q and R/S is defined by (⊕):

$$\frac{P}{Q} \oplus \frac{R}{S} = \frac{P+R}{Q+S}$$

Thus if we consider that our two basic waveforms are the 1^0 and the $^0/_1 \oplus {}^1/_1 = {}^1/_2$, corresponding to a 1^1, the one large-one small pattern. Between this and the 0^1, the Farey sum is $^2/_3$, or one large-two small. Continuing in this way, we generate all the steps in the staircase. The sequence of mixed-mode patterns formed by taking these Farey sums from the 1^0 and 0^1 waveforms is called a 'concatenation' of these states.

There is yet further detail. Examining the border between any two of the mixed-mode forms reveals more concatenations. Taking 1^2 and 1^3 states, for example, researchers have found the following patterns between them (in order of increasing flow rate): 1^2; $(1^2)^21^3$; 1^21^3; $1^2(1^3)^2$; $1^2(1^3)^3$; $1^2(1^3)^4$; $1^2(1^3)^5$; 1^3. The notation $(1^2)^21^3$ means one large-two small, one large-two small, one large-three small per complete cycle. The firing numbers are, respectively, $^2/_3$, $^7/_{10}$, $^5/_7$, $^8/_{11}$, $^{11}/_{15}$, $^{14}/_{19}$, $^{17}/_{23}$, and $^3/_4$. This sequence follows the rules of Farey arithmetic and gives a smaller Devil's staircase between the steps at $N = {}^2/_3$ and $^3/_4$ in Figure 9.3

Similar sequences are found between each of the main treads of the staircase, with the oscillations becoming increasingly more complex at every level of magnification. More than this, some experiments show that the oscillations can become so complex that they lose their periodicity. Such aperiodic, or chaotic, responses have no repeating unit.

chemistry. No experiment is perfectly controlled; the inefficient mixing might be causing the chaos. The pumps used in flow reactors impose a small oscillatory pulse of their own, which we can minimize but not eliminate.

The chemical mechanism for the B-Z reaction is now well established. Richard Field, Endre Körös and Richard Noyes at the University of Oregon developed a scheme in 1974, now called the Oregonator scheme, which has explained the details of the reaction very well. Modelling the reaction on a computer, however, has raised doubts – at least in the minds of some chemists – about whether the B-Z reaction can show genuine chaotic dynamics. Computations for the flow systems have reproduced many of the most complex patterns seen in experiments, but they do not predict chaos convincingly. Some chemists have suggested that the reaction is always trying to be periodic, but that external factors turn this into 'noise'.

Researchers are hotly debating chaos in the B-Z system. Different research groups are re-running each other's computations, often not just drawing different conclusions but obtaining completely different results. At this stage, it is probably wisest to go back to the chemistry of combustion and see if we can find chaos there.

As I hinted earlier, combustion can produce oscillations in reaction rate. The simplest combustion reaction is the oxidation of hydrogen to produce water. At low pressures and temperatures of between 700 and 800 K, mixtures of hydrogen and oxygen ignite spontaneously. In a batch system, such as a closed bulb, this ignition is a one-time event because all the fuel is consumed. In a flow system, there is a fresh supply of reactants. After an ignition, the water formed is pumped out and the vessels refilled with hydrogen and oxygen, and so the process can start again. In this way, the ignition starts to oscillate. The water formed inhibits the reaction, so providing negative feedback and preventing a steady flame from being established. At the lowest temperature in this range, the oscillations have a large amplitude and a long period. At higher temperatures, we encounter oscillations with a small amplitude and a high frequency. As with the B-Z reaction, the transition between the two kinds of oscillation at intermediate temperatures can give rise to a complex sequence of mixed-mode patterns. We still do not know if any of these patterns are chaotic, but many of the warning signs are there.

Oscillations are also common in the burning of the most familiar fuels, the hydrocarbons. An important intermediate in the oxidation is acetaldehyde (ethanal, CH_3CHO). This shows complex ignitions that can lead to engine 'knock' in cars. Computer modelling suggests that the dynamics are extremely sensitive to the initial conditions of the experiment, with chaos appearing when the temperature of the experiment is changed by only a fraction of a degree. At the moment we cannot test these predictions because they are beyond the limits of control in equipment available today.

The dynamic subtleties of such industrially significant reactions can have economic consequences. The chemical industry uses metals as catalysts in many important processes, for example; catalysts can speed up a chemical reaction many thousandfold. Using them efficiently can save a lot of money. Specific sites on the metal's crystalline surface act as templates for the reacting molecules, bringing them close together, so lowering the energy of the reaction and making it happen more easily. Usually the catalyst works better when prepared as a fine powder of tiny crystals, say between 4 and 10 nanometres across (a nanometre is a millionth of a millimetre). When spread evenly onto an inert support, such as zeolite, they provide the maximum surface area on which to catalyse the reaction.

The interesting thing about this set-up is that although the crystals are 'separate from each other', they do not behave independently. How one metal site behaves affects those close to it. The metal sites can exist in either of two different crystalline forms: one is chemically active so can catalyse the reaction, the other is not. When the catalyst is working, each site repeatedly changes from one form to another.

The rate at which this change happens at any one site depends on the state of its neighbours. It may be that the heat produced by a reaction at an active site passes on to a neighbouring inactive one, and there causes a 'phase transition' to the active form. As you might expect, the reaction rate for the whole catalyst depends at any moment on the number of active sites operating.

It also won't surprise you to learn that this kind of cooperative behaviour can lead to chaotic dynamics. To try and model what is going on, we use a computer technique called cellular automata. This approach, where a simple set of rules model evolving complex behaviour, is used in many areas of research dealing with cooperative but potentially chaotic systems.

To see how it works, think of a chessboard with the squares representing individual catalytic sites. A given square is coupled to its nearest neighbours – the four squares sharing an edge with it. We represent the state of the site – active or inactive – by a whole number, i. If i is zero, the site is inactive, otherwise it is active. To play the cellular automaton 'game', we must specify a set of rules that determine how i for each square changes from one moment to the next. This means we have to establish how the rate at which sites change from one state to another depends on what proportion of sites are active or inactive. When the value of i reaches some predetermined value, the site becomes inactive at the time step, with i set equal to zero.

Figure 9.4 shows some of the possible responses of the automaton. At certain values i, the fraction of active sites oscillates in a distinctly irregular way that is reminiscent of what we find experimentally. In other cases, the waveforms become much simpler. The fractional activity of the catalyst is one feature of the system, but we can also wonder how that activity is distributed around the zeolite chessboard. This turns out to be neither random nor uniform. Instead, distinct spatial patterns can emerge. If we ascribe a colour to the value of i on each site, then we get the spectacular spirals shown in Plate 15. The rings propagate outwards with time, giving 'target' patterns, and the spirals unwind. We see similar patterns in the unstirred B-Z mixtures.

Oscillations and chaos in inorganic or gas-phase chemical reactions are interesting but it would be even more exciting to

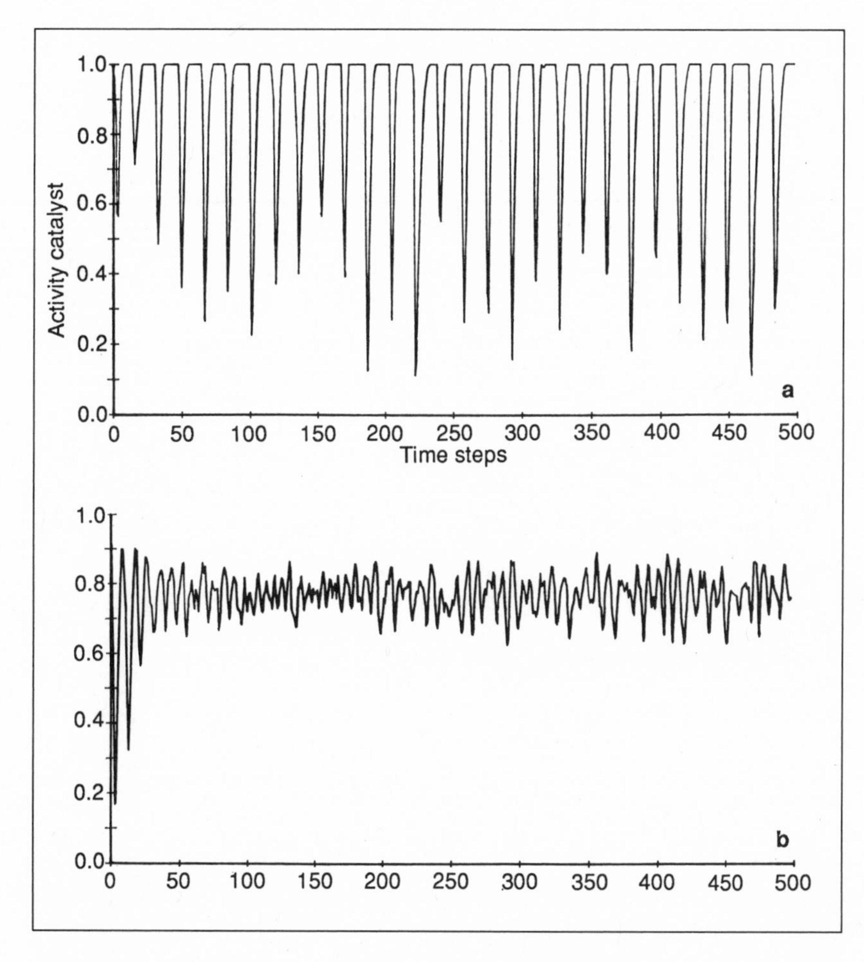

Figure 9.4 Oscillations for cellular automata modelled for some catalysts. In **a**, they couple to give a fairly regular periodic behaviour; in **b**, there are chaotic variations around an average level of catalyst activity.

find chaos in biochemical systems. As Robert May described in Chapter 7, there are many natural rhythms in living organisms, the heartbeat and the 24-hour 'body clock' of circadian rhythms, and so on, where feedback is important, so giving the potential for chaos. So we would expect to find similar behaviour in the underlying biochemistry.

A well-established and much studied oscillation is that in the glycolytic pathway for digesting glucose. Part of this process involves repeatedly adding and removing hydrogen from an important 'coenzyme' involved in the reaction – nicotinamide adenine dinucleotide, mercifully known as NAD, or NADH when hydrogenated. A standard way of detecting NADH is to tag it with a chemical marker that fluoresces in ultraviolet light. In extracts from yeast cells, the emission of fluorescent light oscillates with a period of about five minutes. Biochemists understand the mechanism quite well and have modelled it mathematically. The process involves an 'allosteric' enzyme, which consists of several subunits, each having its own 'active site' where a particular substrate molecule binds and then reacts. The binding of a substrate at one site may influence what happens at others, either activating their ability to catalyse a chemical reaction or inhibiting it. In glycolysis, one of the molecules produced by the reaction activates the sites.

Biochemical oscillators also appear in the nervous system. Transmission in nerves is a very complex process and there is a lot of scope for feedback mechanisms. Take one important reaction involved in the so-called secondary messenger network. In neurons, calcium ions combine with a protein called calmodulin. At low intracellular concentrations of calcium, the ions combine with calmodulin to form a complex that goes to produce a molecule called cyclic-AMP. This causes the calcium ions to be released from the complex. Eventually, at high concentrations of free calcium ions, the enzyme stops working. The same concentration of calcium ions can either activate or inhibit the reaction, depending on the system's previous history.

Although biochemists are very aware of these kinds of complex feedback mechanisms, many chemists find it difficult to accept that even quite simple autocatalytic reactions can show chaotic behaviour. Chemists nearly always assume that reactions proceed under conditions of thermodynamic equilibrium directly towards a state where the products of the reaction have a lower energy than the starting materials. They do not even expect concentrations of intermediate products to oscillate back and forth. But as the Nobel laureate Ilya Prigogine, from the Institute of Chemical Physics in Brussels, pointed out in 1968, many reactions happen in conditions far from thermodynamic equilibrium, producing the kind of behaviour that Belousov first noticed. Prigogine and his group were some of the first chemists to publicize the connection between chaos and nonequilibrium thermodynamics.

Living systems are certainly far from equilibrium, so we should certainly expect to find plenty of chaotic chemistry there. Chemists, especially those who study the details of organic and inorganic biochemical reactions, need to shake off some of their traditional views and accept that, as in other areas of science, there are also limits to predictability in chemical dynamics.

Further reading

R. J. FIELD and M. BURGER (eds.), *Oscillations and Travelling Waves in Chemical Systems*, Wiley, 1985.

AGNESSA BABLOYANTZ, *Molecules, Dynamics and Life*, Wiley, 1986.

A. V. HOLDEN (ed.), *Chaos*, Manchester University Press, 1986.

P. GRAY and S. K. SCOTT, *Oscillations and Instabilities*, Oxford University Press, 1990.

S. K. SCOTT, *Chemical Chaos*, Oxford University Press, 1991.

10
Fractals – a geometry of nature

BENOIT MANDELBROT

Fractal geometry plays two roles. It is the geometry of deterministic chaos and it can also describe the geometry of mountains, clouds and galaxies.

Science and geometry have always progressed hand in hand. In the 17th century, Johannes Kepler found that he could represent the orbits of the planets around the Sun by ellipses. This stimulated Isaac Newton to explain these elliptical orbits as following from the law of gravity. Similarly, the back-and-forth motion of a perfect pendulum is represented by a sine wave. Simple dynamics used to be associated with simple geometrical shapes. This kind of mathematical picture implies a smooth relationship between an object's form and the forces acting on it. In the examples of the planets and the pendulum, it also implies that the physics is deterministic, meaning that you can predict the future of these systems from their past.

Two recent developments have deeply affected the relationship between geometry and physics, however. The first comes from the recognition that nature is full of something called deterministic chaos. There are many apparently simple physical systems in the Universe that obey deterministic laws but nevertheless behave unpredictably. A pendulum acting under two forces, for example. The notion of deterministic yet unpredictable motion is a surprise to most people.

The second development came from efforts to find mathematical descriptions for some of the most irregular and complicated phenomena we see around us: the shapes of mountains and clouds, how galaxies are distributed in the Universe, and examples nearer home. One way of obtaining such a description is to seek a 'model'. In other words, I had to invent or identify mathematical rules that can produce 'mechanical forgeries' of some part of the reality – a photograph of a mountain or a cloud, a map of deepest space, or a chart on the financial pages of a newspaper.

Indeed, Galileo proclaimed that 'the great book of nature is written in mathematical language', adding that 'its characters are triangles, circles and other geometrical figures, without which one wanders in vain through a dark labyrinth'. Such Euclidean shapes have, however, proved to be quite useless in modelling either deterministic chaos, or irregular systems. These phenomena need geometries that are very far from triangles and circles. They require non-Euclidean structures – in particular, a new geometry called fractal geometry.

I coined the word fractal in 1975 from the Latin *fractus*, which describes a broken stone – broken up and irregular. Fractals are geometrical shapes that, contrary to those of Euclid, are not regular at all. First, they are irregular all over. Secondly, they have the same degree of irregularity on all scales. A fractal object

Figure 10.1 This cauliflower, a variety called *c. Romanesco*, is an example of a natural fractal.

looks the same when examined from far away or nearby – it is self-similar. As you approach it, however, you find that small pieces of the whole, which seemed from a distance to be formless blobs, become well-defined objects whose shape is roughly that of the previously examined whole.

Nature provides many examples of fractals, for example, ferns, cauliflowers and broccoli, and many other plants, because each branch and twig is very like the whole. The rules governing growth ensure that small-scale features become translated into large-scale ones.

A striking mathematical model of the way fractals work is the Sierpiński gasket. Take a black triangle and divide it into four smaller triangles, as shown in Figure 10.2, and erase the central

Figure 10.2 The Sierpiński gasket – a simple fractal produced by breaking up a triangle into successively smaller ones.

fourth triangle so that it leaves a white triangle. Each new black triangle will have sides that are half as long as the initial triangle. Repeat the exercise with each new triangle and you obtain the same structure on an ever decreasing scale with a detail that is twice as fine as that in the preceding stage. When parts of the object are exactly like the whole, the object is said to be linearly self-similar.

However, the most important fractals deviate from linear self-similarity. Some of these are fractals that are generated by a random process, while others are fractals that can describe chaotic, or nonlinear, systems (where the factors affecting the way the system behaves are not proportional to the effects they produce). Let us take one example of each.

Random fractals became best known through the stream of forgeries of coastlines, mountains and clouds, such as the one in Plate 17, which my colleagues and I have been producing since 1975 using computer graphics. Other examples are some of the scenes made for films such as *Star Trek II*.

Our work on such fractal modelling began with a bit of folk wisdom and a lot of natural history. The folk wisdom started with observing something that even a cubist painter knows. 'Clouds are not spheres, mountains are not cones, coastlines are not circles, and bark is not smooth, nor does lightning travel in a straight line.' All of these natural structures have irregular shapes that are self-similar. In other words, we discovered that successively magnifying a part of the whole reveals a further structure that is nearly a copy of the original we started with.

The natural history involved collecting and classifying facts about natural structures. For example, as you measure the coast of a country with ever increasing precision, its length becomes greater because you have to take into account ever smaller irregularities along the length. Lewis Fry Richardson has found an empirical law that describes this increase.

To make sense of fractal geometry we have to find ways of expressing the shape and complexity in terms of numbers, just as Euclidean geometry uses the notions of angle, length, area or curvature, and the notions of one, two or three dimensions.

For complicated geometrical objects, the ordinary notion of dimension may vary with scale. As an example, take a ball with a diameter of 10 centimetres made of a thread of 1 millimetre. From far away, the ball appears as a point. From a distance of 10 centimetres, the ball of thread is three-dimensional. At 10 millimetres, it is a mess of one-dimensional threads. At 1 millimetre, each thread becomes a column and the whole becomes a three-dimensional object again. At 0·1 millimetres, each column dissolves into fibres, and the ball again becomes one-dimensional, and so on, with the dimension 'crossing over' repeatedly from one value to another. When the ball is represented by a finite number of atom-like pinpoints, it becomes zero-dimensional again.

For fractals, the counterparts of the familiar dimensions (0, 1, 2, 3) are known as fractal dimensions. Usually, their values are not whole numbers.

The simplest variant of fractal dimension is the similarity dimension D_s. Applied to a point, a line, a square or a cube, D_s simply gives the number of ordinary dimensions needed to describe the object – 0, 1, 2, 3 respectively. What about a curve that is a linearly self-similar fractal? Such a curve can range from being an almost smooth, one-dimensional line to being nearly plane filling, which means that the line twists and turns so much that it visits nearly every part of some region of the plane, becoming almost two-dimensional. Correspondingly the value of D_s will range up from just above 1 to just below 2. Thus, D_s can be said to measure the complexity of this curve. More generally, D_s measures the complexity or degree of roughness of a fractal shape.

Another simple fractal dimension is the mass dimension. The mass in a one-dimensional straight rod increases in proportion to its length, say $2R$. The mass in a two-dimensional disc of radius R increases in proportion to πR^2, the area of a circle. And the mass in a ball increases in proportion to $\frac{4}{3}\pi R^3$, the volume of a sphere. So when a further dimension is added, the mass grows in proportion to R raised to a power indicating the number of dimensions.

In the case of a fractal, the mass grows proportionately to R raised to some power D_m that is not a whole number. That is, D_m plays one of the usual roles of dimension, so it is natural to call it a fractal dimension. It is fortunate that in all the simple cases, D_s and D_m (and other definitions of fractal dimension) take precisely the same value. In cases beyond the simplest they may differ.

The next step in modelling is to imagine the simplest geometrical construction that might have the right properties to generate the structure. In fact, I have put together, and constantly enrich, a toolbox of such constructions that are useful for fractal geometry. To test which mathematical tool would be appropriate, we compare the numerical characteristics of the model with the real thing – the fractal dimensions of a mountain, for example. This is not enough, however. We also use computer graphics to test how good a tool we have.

At the end of the day, we hope to produce a theory from the fractal modelling of mountains that can describe the relief of the Earth.

Because fractals have proved useful in describing complex natural shapes, it is not surprising that fractals also play a part in describing how complex dynamical systems behave. As previous chapters have shown, the equations that model turbulence in liquids, the weather, or the dynamics of insect populations are nonlinear and show behaviour typical of deterministic chaos. If we iterate these equations – examine their solutions as they evolve over time – we find that many of the mathematical properties, especially when shown as computer graphics, reveal themselves to be self-similar. Examples are the 'phase portraits' of so-called strange attractors described in the chapter by Ian Stewart.

My best-known contribution to this area of nonlinear fractals is called the Mandelbrot set (see 'Order and chaos in the Mandelbrot and Julia sets', below). The set results from iterating a relatively simple equation. It produces the most extraordinary graphics, rich in complexity. Some people have called it the icon for nonlinear fractal geometry.

Order and chaos in the Mandelbrot and Julia sets

Take a starting point C_0 in the plane of coordinates u_0 and v_0. From the coordinate of C_0, form a second point C_1 of coordinates $u_1 = u_0^2 - v_0^2 + u_0$ and $v_1 = 2u_0v_0 + v_0$. Next, form the point C_2 of coordinates $u_2 = u_1^2 - v_1^2 + u_0$ and $v_2 = 2u_1v_1 + v_0$. More generally, the coordinates u_k and v_k of C_k are obtained from u_{k-1} and v_{k-1} by the so-called iterative formulas $u_k = u^2_{k-1} - v^2_{k-1} + u_0$ and $v_k = 2u_{k-1}v_{k-1} + v_0$.

These formulas may seem a little artificial, but they simplify further if expressed in terms of the mathematics where these formulas originate. A point C of coordinates u and v is represented by the complex number $c = u + iv$. One can add and multiply complex numbers, except that *i^2 must always be replaced by -1. When* C_0 is represented $c_0 = u_0 + iv_0$, it is easy to verify that the above formulas simply express that $c_1 = c_0^2 + c_0$ and $c_k = c^2_{k-1} + c_0$. Even the reader who is scared of complex numbers understands the expression in terms of u_k and v_k.

The points C_k are said to form the orbit of C_0. The Mandelbrot set (M) is defined as follows: if the orbit C_k fails to go to infinity, the point C_0 is said to be contained within the set M; if the orbit goes to infinity, then the point C_0 is outside M.

The original reason for writing this algorithm is that it concerns the problem of the so-called quadratic complex dynamics. When C_0 is in the interior of M, the dynamics yields an orbit that is perfectly orderly. When C_0 is outside M, the behaviour of the orbit is deterministic but practically unpredictable, or chaotic.

The boundary between orderly and chaotic behaviour turns out to be unbelievably messy, so that the set M combines the two themes of order and chaos in more than one way. For example, Plate 19 represents a zoom upon a tiny piece of M, reduced in ratio of 10^{-23} to 1. The little 'bug' near the centre has very nearly the same shape as the whole Mandelbrot set (this is a token of geometric orderliness). On the other hand, the surrounding pattern is not present in the whole set, but depends very much upon the point on which the zoom has focused (this is a token of variety, and even chaos).

By changing the rules that produce the Mandelbrot set, one obtains a shape called the Julia set. As before, you pick a point C of coordinates u and v, and now you call it a 'parameter'. Next you pick, in a different plane, a point P_0 of coordinates x_0 and y_0. Then you form $x_1 = x^2_0 - y^2_0 + u$ and $y_1 = 2x_0y_0 + v$. Compared with the rule for the Mandelbrot set, this rule mixes the points (x_0, y_0) and (u, v). In terms of the complex numbers $c = u + iv$ and $z = x + iy$, the rule is $z_1 = z^2_0 + c$ and (more generally) $z_k = z^2_k + c$.

When the orbit P_k fails to escape to infinity, the initial P_0 is said to belong to the 'filled-in Julia set'. To prepare Plate 20, Alan Norton went a step further, replacing the complex number z, which is a point in the plane, by a quaternion, which is a point in four-dimensional space. Two difficulties arose: the fourth dimension cannot be plotted. and the Julia set would be so bulky that its structure would be unclear. Therefore, the illustration takes advantage of the fact that there are several 'components' to this Julia set, and represents only two of them, in different colours.

The Mandelbrot set does not just produce beautiful pictures. If we examine many pictures with great care, we find innumerable empirical observations that can be restated in the form of mathematical conjectures. Many of these have already led to brilliant theorems and proofs. It has also inspired a new approach to mathematics, using a computer screen.

Mathematical conjectures usually originate in previously known theorems. In recent decades, there was no input at all from physics or from graphics, which meant that some areas of pure mathematics, such as the theory of iteration (to which the Mandelbrot set belongs) ran out of steam. Fractal pictures done on the computer have revived it. Being able to play with pictures interactively has provided a deep well for mathematical discoveries. Examining the Mandelbrot set has led to many conjectures that were simple to state but hard to prove.

Of course, many related fractals lead to beautiful and intriguing graphics. Indeed, several shapes known today as fractals were discovered many years ago. Some of these mathematical entities appear in the work of the French group of mathematicians Henri Poincaré, Pierre Fatou and Gaston Julia, at various times from 1875 to 1925. But nobody realized their significance as visual descriptive tools and their relevance to the physics of the real world.

One model where random fractals describe the real world is a form of random growth called diffusion limited aggregation, or DLA (see Figure 10.3). This yields tree-like shapes of baffling complexity. DLA can be used to model how ash forms, how water seeps through rock, how cracks spread in a solid and how the lightning discharges.

To see how it works, take a very large chess board and put a queen, which is not allowed to move, in one of the central squares. Pawns, which are allowed to move in any of the four directions on the board, are released from a random starting point at the edge of the board, and are instructed to perform a random, or drunkard's, walk. The direction of each step is chosen from four equal probabilities. When a pawn reaches a square next to that of the original queen, it transforms itself into a new queen and cannot move any further. Eventually, a branched, rather spidery-looking collection of queens, a "Witten Sander DLA cluster', grows.

Quite unexpectedly, massive computer simulations have shown that DLA clusters are fractal; they are nearly self-similar. Small portions are very much like reduced versions of large portions. But clusters deviate from randomized linear self-similarity, something that will pose interesting challenges for the future.

What is special about this kind of fractal growth is that it shows very clearly how parameters that vary smoothly produce rugged behaviour. To show how, let us rephrase the original construction in terms of the theory of electrostatic potential. Imagine a big box, in which the DLA is constructed, is set to a positive electrical potential, and the target object, the original

Figure 10.3 A random fractal called a 'diffusion limited aggregate' produces fern-like shapes that model lightning and other natural phenomena.

queen, is put at the centre and set to a potential of zero. What is the value of the potential elsewhere in the box?

In the cases where the outline of the central object is a smooth curve, or has small number of kinks, like a triangle or a square, scientists have long known how to compute the potential. These classical analytical calculations determine the curves along which the potential is the same. All these curves are smooth, and they provide progressive transitions between the fixed box and the boundary of the fixed object at the centre. Next, suppose that the boundary includes a needle-like projection. Around the needle, the equipotential curves would be very crowded. The drop in potential would be very steep, causing an electrical charge: the needle would act like a lightning rod. When the central object is a DLA cluster, all its boundary is lined with needles, and lightning will fall most often on those needles that are the most exposed.

Here comes a critical novelty: the mechanism of DLA is equivalent to postulating that, after a needle has been struck by lightning, it extends or branches. The experiments on DLA teach us that, when we allow boundaries to move in response to the potential, the cluster grows into an increasingly large DLA structure. This implies we can create rough fractals from the smoothness characteristic of the equation generating the lines of equal potential. Thus, in this context, fractal geometry has led to a new problem and a new area of investigation.

Fractal geometry is also being used to describe many other complex phenomena in nature. One of the most fruitful areas is in studying turbulent motion, not only how it arises – the dynamics when displayed as a phase portrait are fractal – but also the complex shapes of turbulent structures. Thus, wakes and jets of water and clouds turn out to be fractal. This must be due to the action of the equations of fluid motion – the Navier-Stokes equations. The problem of relating shape to the dynamics producing it, however, is still wide open. Tracing this relationship would be a major step in understanding turbulence.

Another area where fractals provide an apt description are in living things and in the Universe at large, although in each case the fractal description breaks down on the very small scale and on the very large. Trees or arteries do not branch endlessly, and whole trees are not part of supertrees. The opposite may be true for the distribution of galaxies in the Universe. Counts of galaxies yield undisputed evidence that at comparatively small scales the distribution is fractal. These small scales are known to extend at least as far as 5 to 10 megaparsecs. There is increasingly strong evidence that there are large voids at a size well above 100 megaparsecs. Such voids are precisely what is expected in a fractal distribution.

How important are fractals? Like for the theory of chaos, it is too early to say for sure, but the prospects are favourable. Many fractals have already had an important cultural impact and have already been accepted as works of a new form of art. Some are representational while others are totally unreal and abstract. It must come as a surprise to both mathematicians and artists to see this kind of cultural interaction.

To the layman, fractal art seems to be magical. But no mathematician can fail to try to understand its structure and its meaning. Much of the underlying equations would have been regarded as part of being pure mathematics, without any application to the real world, had its visual nature not been seen.

Multifractal measures

If we take a simple geometrical shape such as the Sierpiński gasket, we can make some simple and useful mathematical observations. Given a point *P* in a plane, the question: 'Does *P* belong to the Sierpiński gasket?' has a black-and-white answer, 'yes' or 'no'. Such statements define the gasket as being a fractal set.

Most questions, however, particularly in nature, have a more complicated answer. For example, 'Is there any oil in the south of England?' calls for a 'many shades of grey' answer, ranging from 'bare traces' to 'enough to pump even when the price of oil falls very low'.

This last answer applies to only

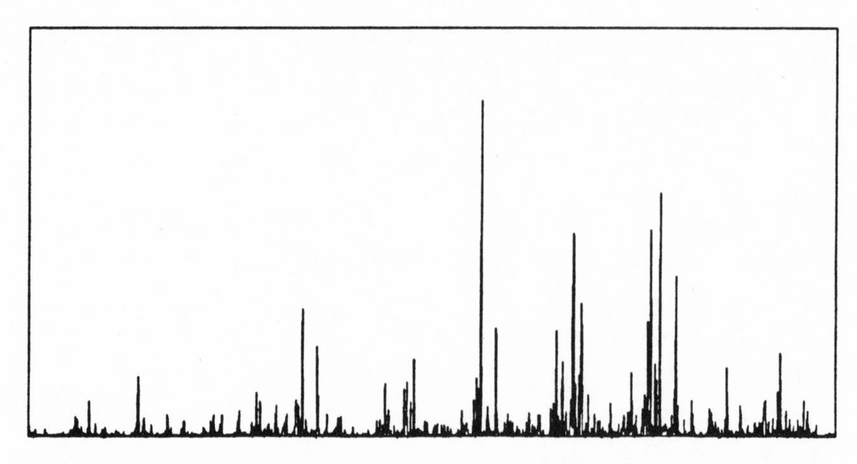

Figure 10.4 Here is a typical trace of multifractals, in this case modelling turbulence in a laboratory experiment, which has a similar structure to multifractals describing oil distribution.

a few places, which are unevenly distributed, because they are clustered in few parts of the world. Furthermore, even these parts are mostly barren, so that the distribution of oil is just about as irregular within a country as it is elsewhere in the world. This is an extension of the notion of self-similarity. If we take a line around the globe and plot the availability of oil, we obtain a very wiggly line, with a few sharp peaks separated by large regions of flatness.

How do we represent such an irregular pattern? If we divide the world into regions, each characterized by a different level of oil resources, it happens that each of these regions is nearly a fractal set. So the pattern is not itself a fractal set, but a novel combination of a multitude of fractal sets. It is called a multifractal distribution, or multifractal measure.

Most important, as already mentioned, many of the most active uses of fractals are in physics, where they have helped tackle some very old problems and also some altogether new and difficult ones.

A final satisfying spin-off from fractal pictures is that their attractiveness seems to appeal to the young and is having an influence on restoring interest in science. Many people hope that the Mandelbrot set and other fractal pictures, now appearing on T-shirts and posters, will help to give the young a feeling for the beauty and eloquence of mathematics, and its profound relationship with the real world.

Further reading

B. B. MANDELBROT. *The Fractal Geometry of Nature*, W.H. Freeman, 1982.

H. O. PEITGEN and P. H. RICHTER, *The Beauty of Fractals*, Springer Verlag, 1986.

J. FEDER, *Fractals*, Plenum, 1988.

11
Fractals, reflections and distortions

CAROLINE SERIES

Fractals obtained from repeated reflections in circular mirrors produce breathtaking kaleidoscopic images. Understanding these pictures may give us new insights into the geometry of chaos.

Most people have become familiar in recent years with pictures of fractals, those elusive shapes that, no matter how you magnify them, still look infinitely crinkled. The pictures you saw were probably drawn by computer, but examples abound in nature – the edge of a leaf, the outline of a tree, or the course of a river. Fractal curves differ from those studied in normal geometry. The curve of a circle, for instance, if magnified sufficiently, just about becomes a straight line. A fractal curve, on the other hand, when viewed on many different scales, from macroscopic to microscopic, reveals the same intricate pattern of convolutions. How do you construct a fractal curve? A simple example is the famous Koch snowflake, invented by Helge von Koch in 1904. It is an example of a 'nowhere smooth' curve.

To draw the snowflake, start with the triangle shown in Figure 11.1a. Then replace each of the sides of this triangle by a bent line as shown in Figure 11.1b. At the next stage, Figure 11.1c, each of these sides in turn is replaced by the same pattern but on a smaller scale, and so on, ad infinitum, to obtain finally the snowflake shown in Figure 11.1d.

In recent years, there has been a revolution of interest in fractals. Previously, only a few people had appreciated the significance and beauty of these strange shapes. Benoit Mandelbrot

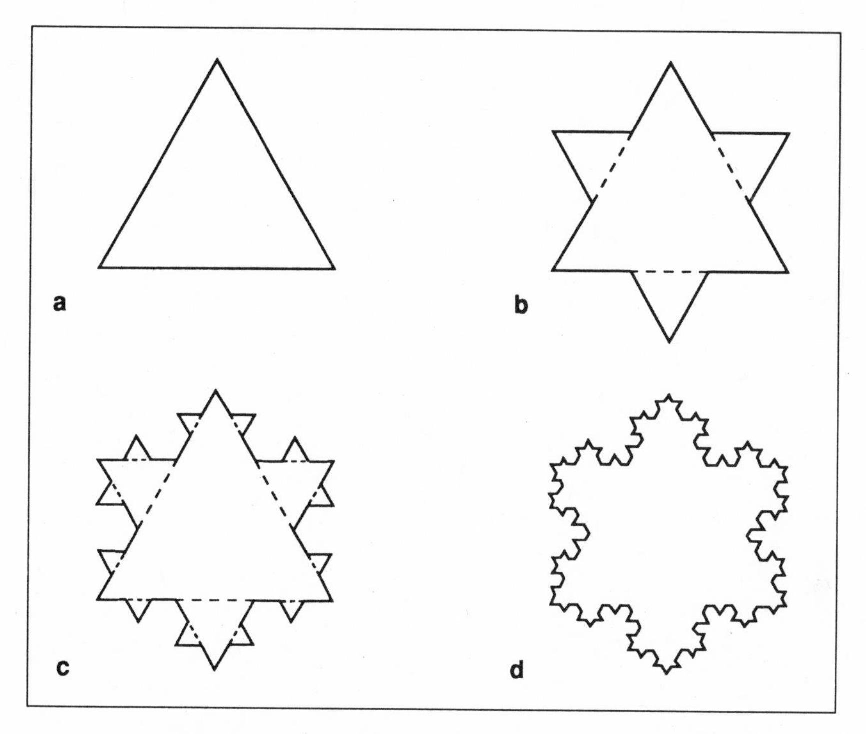

Figure 11.1 The Koch snowflake is a fractal evolving from a simple triangle.

drew much attention to their potential use in describing the natural world. At the same time, the development of high-speed computing and computer graphics has made them easily accessible and this has drawn many people to study them more closely.

Another reason for the interest in fractals is that they are connected with chaos. In mathematics, chaos has a specialized meaning. The easiest way to understand chaos is by some examples.

Suppose a particle is moving in a confined region of space according to a definite deterministic law. Following the path traced out by the particle, we are likely to observe that it settles down to one of three possible behaviours – the geometrical description of which is called an attractor. The particle may be attracted to a final resting position (like, for example, the bob on pendulum as it gradually settles down to rest). In this case, the attractor is just a point – the final resting position of the bob. Or the particle may settle down in a periodic cycle (like the planets in their orbits around the Sun). Here the attractor is an ellipse and the future motion can be predicted with astonishingly high accuracy as far ahead as we want. The last possibility is that the particle may continue to move wildly and erratically while, nevertheless, remaining in some bounded region of space. The motion of some of the asteroids, for example, appears to exhibit exactly this phenomenon. Tiny inaccuracies in measuring the position and speed of the asteroid quickly lead to enormous errors in predicting its future path. This phenomenon is the signal of chaotic motion. The regions of space traced out by such motions are called strange attractors.

Once a particle is attracted to a strange attractor there is no escaping. Almost anywhere you start inside the attractor, the point moves, on the average, in the same way, just as no matter how you start off a pendulum, it always eventually comes to rest at the same point. Although the motion is specified by precise laws, for all practical purposes, the particle behaves as if it were moving randomly. The interesting point here is that strange attractors are very frequently fractals.

You might think that to generate such complicated behaviour, the equations governing the motion would have to be very complicated. An important insight of recent years has been that, on the contrary, chaotic motion frequently follows simple deterministic laws.

Mathematicians like to study idealizations of this kind of phenomenon. They have found a fertile ground of investigation in what is called the iteration of polynomials. Polynomials are expressions containing one or more variables raised to some powers, such as $w = z^2 + 1$. Iteration is a process whereby you take such an expression, feed in a particular value for the variable, z, calculate the answer, w, which then becomes the new value of z. The operation is repeated over and over again. Iteration is a way of modelling deterministic behaviour – each new position of a moving particle, as represented by the point z, depends on its previous position.

In the simplest example of iterating polynomials, you take a point z and let it evolve over a series of steps so that the new value of z is the square of the previous value of z plus a constant. This is expressed as $z \rightarrow z^2 + c$ where c is a constant. If z and c are complex numbers, we obtain some interesting results. A complex number contains two variables x and y, where y is multiplied by an imaginary number, the square root of minus one (see 'Practical uses for imaginary numbers', below). So z is written as $x + \sqrt{-1}y$. The plane defined by the coordinates x and y, in which z moves as we iterate, is called the complex number plane.

Computer modelling can be used to follow the movement of z. This generates pictures of intricate complexity like those shown in Figure 11.6. These are the strange attractors for this model. They are called Julia sets, after the French mathematician Gaston Julia who, together with his contemporary Pierre Fatou, first studied them in 1918. Despite many beautiful and striking results, this work remained relatively unknown until

Practical uses for imaginary numbers

Complex numbers are numbers of the form $z = x + \sqrt{-1}y$. The number y is called the imaginary part of z. No ordinary number can be the square root of -1, because the square of any ordinary number is always greater than zero. Complex numbers can be added, subtracted and multiplied in the ordinary way, provided only that we remember to substitute -1 for $(\sqrt{-1})^2$ whenever it occurs. The complex number $x + \sqrt{-1}y$ is frequently represented as the point in the plane with coordinates x, y. This is the representation used in drawing pictures of the Julia set. Complex analysis, the calculus of functions of a complex variable, is used daily in the design of aeroplane wings and electronic circuits, in complicated statistical analyses, in quantum theory and in the making and breaking of top secret military codes.

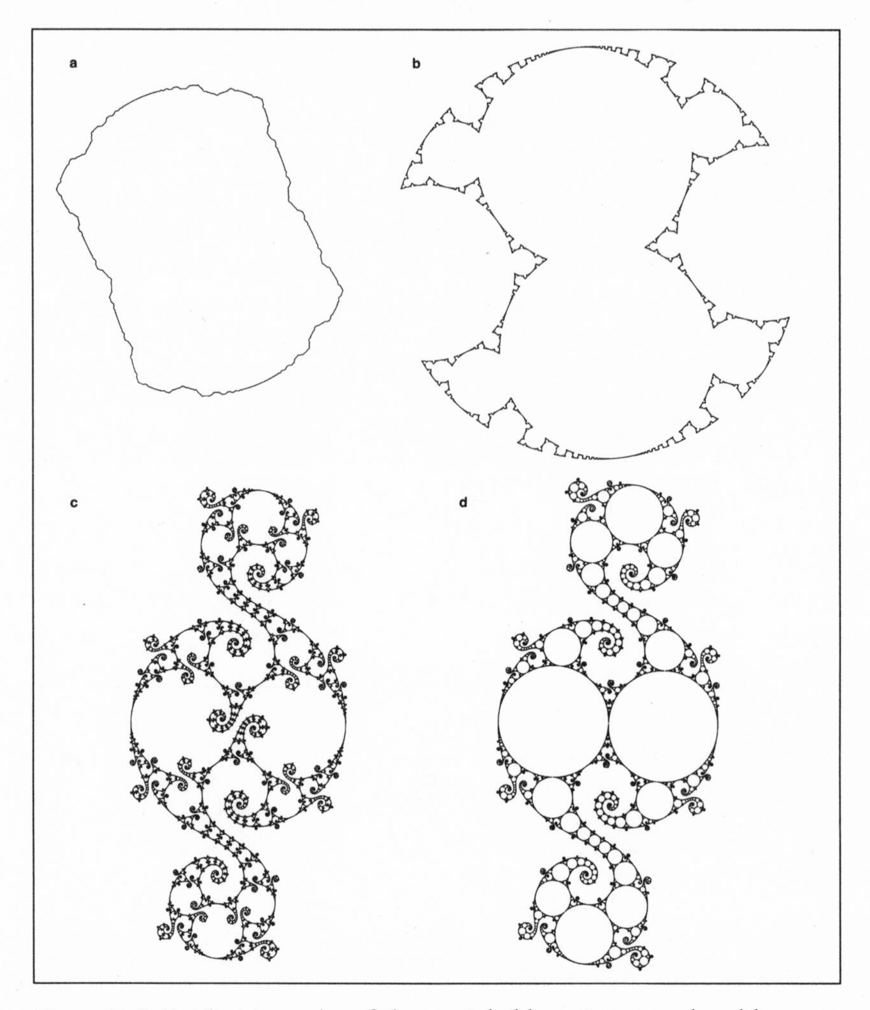

Figure 11.2 Further examples of the remarkable patterns produced by non-Euclidean limit sets.

mathematicians could draw pictures of Julia sets on a computer.

Towards the end of the 19th century, the German mathematician Felix Klein had already studied another kind of set that also merits the name of strange attractor. Klein was investigating non-Euclidean geometry, which had been discovered about 50 years earlier. Klein's fractals arose from iterating not one but several motions of the particle in the complex number plane. These motions were all of a special kind called inversions, which are a sort of reflection. The collection of all permitted inversions, together with those produced by iterating the equations, form what mathematicians call a group. The particular groups that Klein studied are called, despite Klein's strong objections, Kleinian groups. In this case, the strange attractor generated is called a limit set. Start at any point you like in the plane and look at the collection of all its images or 'iterates' under all members of the group. These iterates will always appear to pile up in a certain region of the plane, and this region is the limit set.

Plate 21 and Figure 11.2 illustrate some of the beautiful examples of the limit sets obtained in this way. Plate 21 was drawn by the physicist Roman Tomaschitz, from the University of Brussels, who is interested in these limit sets because they are related to a possible model for what is called quantum chaos. Figures 11.2a–d were drawn at Harvard University by Curt McMullen, David Mumford and David Wright.

In order to understand inversion, think for a moment about reflecting in great circles on a sphere. Such reflections produce the beach ball patterns shown in Figure 11.3a. Figure 11.3b shows the beach ball pattern projected onto a flat surface. Inver-

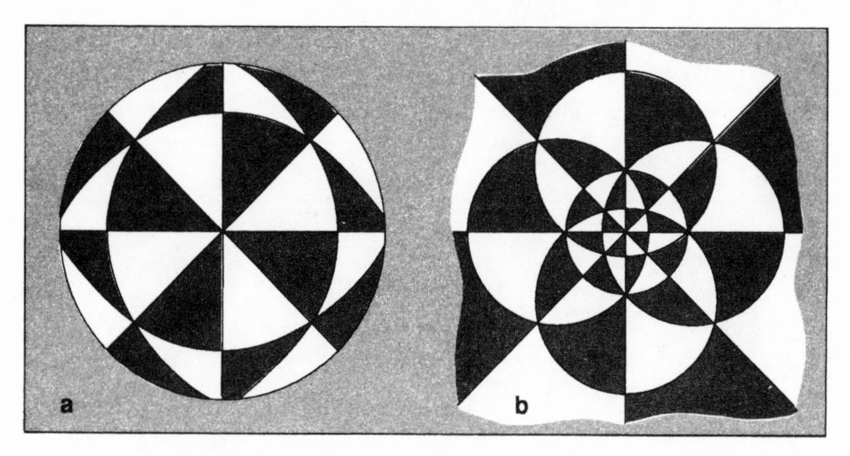

Figure 11.3 Reflecting in great circles on a sphere on to a flat surface, as in **a**, produces the inversion shown in **b**.

sion is nothing other than the effect, as seen on the plane, of reflection in great circles on the sphere. 'Reflections of a globe on a flat map', below, explains it in more detail.

Inversion shares with reflection the property that there is always a 'mirror'. Instead of being a straight line as in the case of an ordinary reflection, the 'mirrors' used for inversions are circles. Inversion interchanges all the points on one side (inside) of this circular mirror with all the points on the other side (outside). Points on the circle itself are left fixed. You can think

Reflections of a globe on a flat map

Figure 11.4 will remind you of how stereographic projection works. Think of the globe as a sphere Σ with the north pole N at the top. The sphere rests on a horizontal plane π, touching π at the south pole S. Stereographic projection transfers figures on the globe onto figures on the plane below. To find the image of a point P' on Σ, join P' to N by a line. Extend this line beyond P' to where it cuts π in Q. The point Q is the projection of P'.

Now reflect Σ in the equatorial plane l, and try to transfer the result to the plane π. The southern hemisphere projects onto the inside of the circle, centred on S, marked k in Figure 11.4. As a result of reflection in l, points inside k are transformed into points outside k and vice versa, while points on k remain fixed.

The diagrams illustrate what happens to a point Q inside k under this transformation. Let P be the point corresponding to Q

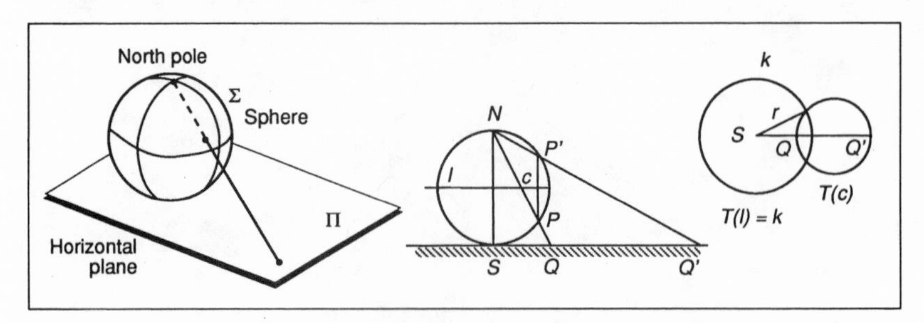

Figure 11.4 How to project a beach ball on to a flat surface.

on Σ, and let P' be the image of P under reflection in l. Let $Q' = T(P')$ be the image of P' under stereographic projection T. In the figure on the far right, you see the circles $T(l) = k$ and $T(c)$ on π. These are the images on π of l and of the circle c through P and P' which is perpendicular to l.

Both l and c, being circles on the horizontal plane π, project to circles on π, and furthermore, since l and c cut at right angles, so do $T(l)$ and $T(c)$. Using some ordinary geometry about circles which cut at right angles, we get the formula $SQ.SQ' = r^2$, where r is the radius of $T(l)$. This is the general formula for inversion in a circle of radius r.

---

of ordinary reflection as a special kind of inversion, in which the circular mirrors have infinite radius. The possibilities you can get from successive inversions in two circles are rather limited. In these cases, the limit set – the region where all the images of any particular point eventually bunch up – consists of either one or two points.

Things begin to get interesting when you start to invert in three or more circles. Figure 11.5 is an example of this. We start

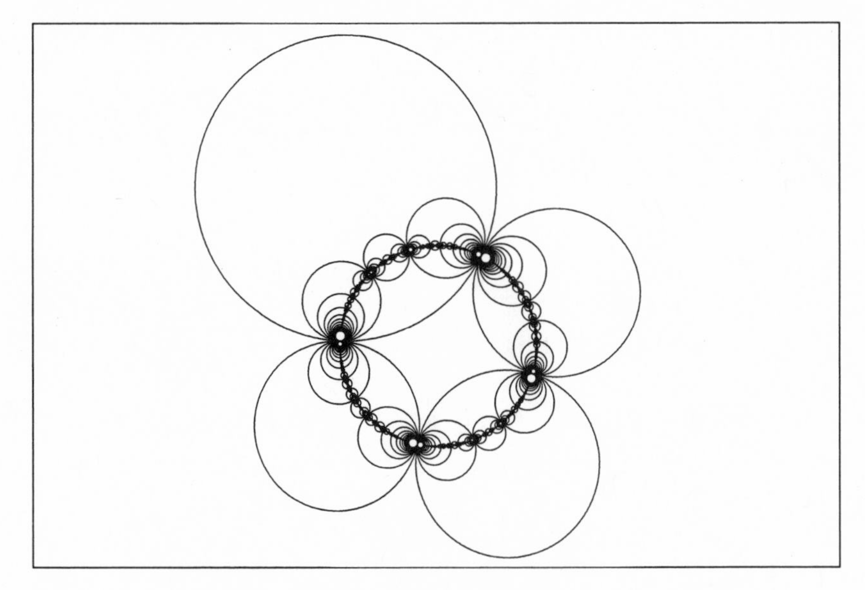

Figure 11.5 shows how inverting circles produces a chain of reflections.

with a base circle and draw four other circles. Each of these circles cuts the base circle at right angles and each touches its two neighbours at the points where they meet the base circle. We then pick any one of these four circles as the 'mirror' for an inversion. The result is that the three other circles invert into three touching circles inside the chosen circle. This is because all four circles meet the base circle at right angles, and because inversion has the property that it transforms circles into circles and leaves the angle of intersection between two circles unchanged.

We do the same thing for the other three circles. At the next stage, we invert in each of these 12 new smaller circles in turn, producing in each one a chain of 11 yet smaller touching circles. And so we continue. It is rather like the chain of reflections you see when the hairdresser holds up a mirror behind you so that you can inspect her or his work.

The picture called 'Les parapluies de Verone' (Plate 22) has been obtained in a similar way, but now there are seven initial circles which intersect near the centre of the picture forming the edge of the central blue umbrella.

Where are the fractals that Klein was studying? So far, fractal curves do not seem to have entered the scene. In fact, the diagrams we have looked at so far are analogous to the iteration of our example $z \to z^2 + c = 0$. In this case, the strange attractor, or Julia set, is nothing other than the circle of unit radius in the plane.

The reason for this is as follows. When $c = 0$, the sequence of iterates of a point z is nothing other than the sequence of powers, z, z^2, z^4, z^8, and so on. If a point starts inside the circle of radius one, after a few iterations, these powers rapidly approach zero. On the other hand, if we start with a point outside this circle, the higher powers of z rapidly get larger and the point goes off to infinity. It is only when we start with a point on the unit circle that its iterates remain always at a distance one from the origin, moving around and round the circle forever.

The amazing fact is that we have only to introduce the smallest possible non-zero value of c and the Julia set gets distorted, not

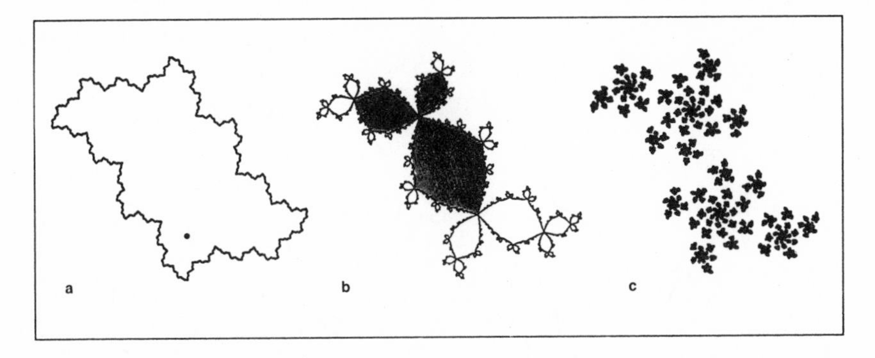

Figure 11.6 shows what happens when we gradually distort the Julia set from a circle. The initial fractal curve in **a** becomes the 'Douady rabbit' in **b**, and then splits into 'islands' in **c**.

into a smooth curve roughly circular in shape, but into a fractal like that shown in Figure 11.6a.

We can introduce a similar kind of distortion into our inversion example in Figure 11.5 by altering the position of the initial circles slightly in such a way that they no longer cut the fundamental circle at right angles. The result is that the fundamental circle becomes distorted into a curve that is a fractal. If we do this in the example of four touching circles, the overall arrangement of small reflected circles remains the same, but the circle no longer passes through their meeting points. Instead, at each level, the small circles form a kind of pearl necklace in which the pearls, instead of lying smoothly alongside each other in a round chain, get more and more jostled out of their original positions. Eventually, the individual beads disappear completely and we are left with the limit set of the distorted group. It is the result of this process that is illustrated in Figure 11.2. The picture shown in Plate 21 is a slightly different example in which some of the initial circles intersect.

What happens as we gradually distort further and further away from the configuration? Some examples of Julia sets obtained in this way are shown in Figures 11.6b and 11.6c. As we distort in the inversion pictures, we begin to get the beautiful Paisley patterns appearing in Figure 11.2.

What happens as we go on distorting? Where does the realm

of distortion end? Sooner or later, it seems, the complication of the curves must become so great that things break down in some way. This is so, and what happens in the two cases is not quite the same.

In the case of $z \to z^2 + c$, the first thing that happens is that certain points of the Julia set come together. This is what happened in Figure 11.6b, which is often known as a Douady rabbit, after the French mathematician Adrien Douady whose work includes some of the most important results in this field. Some time later, the poor rabbit's head, ears, nose, legs and tail are all split off from her body to give the picture in Figure 11.6c. The Julia set has broken apart into a lot of 'islands'. These islands still retain the fractal property: if you look closely, each island is in fact broken up into many smaller islands, and so on ad infinitum. In fact, there is no piece of dry land with any spatial extension at all. A set of this kind is called a Cantor set. We know that strange attractors often take this form.

For inversion groups, the situation is rather different. Total breakdown happens as soon as any points on the 'circular' limit set come together. Last year, Curt McMullen proved that the points where breakdown occurs in this way are spread thickly on the boundary of the distortion region in a manner analogous to that in which recurring decimals are spread thickly among all numbers. Exactly what happens at the remaining points remains a mystery. Beyond this point, things get so bad that the whole plane is filled in by the limit set, so that no interesting pictures can be drawn at all.

Can we describe those parameter values at which the breakdown occurs? By parameter values, we mean the c values, or, in the inversion case, the numbers that describe the initial configuration of circles. It turns out that in the case of four initial circles, we can describe the initial configuration entirely by means of one complex number. Once again, we are in for a surprise. We might legitimately expect the boundary of 'nice' parameter values to be described by a 'reasonable' curve. But not so. Let us look at the situation for $z \to z^2 + c$. What we get is the famous Mandelbrot set illustrated in many places.

The Mandelbrot set is a plot of the set of c values at which the Julia set breaks up into islands to form a Cantor set. It is very important to understand that the Mandelbrot set is not the Julia set of any particular iteration procedure, but the set of points in parameter space where a certain type of behaviour breaks down. The Mandelbrot set has an amazingly intricate structure. The more you magnify the Mandelbrot set near a particular c value the more you see what looks like the Julia set for that same value of c. Extending out of the main body of the Mandelbrot set are a multitude of hairs.

Along each of these hairs, arranged, in the words of Douady, 'like an infinity of droplets on a spider's web in the morning dew', are tiny copies of the Mandelbrot set itself.

What is the analogy of the Mandelbrot set for Kleinian groups? Such a set is called a Teichmüller space. Although mathematicians know a lot about Teichmüller spaces, their actual shapes have not been much studied. A computer-drawn picture for a case similar to the inversion example I have described is shown in Figure 11.7. This picture was drawn by

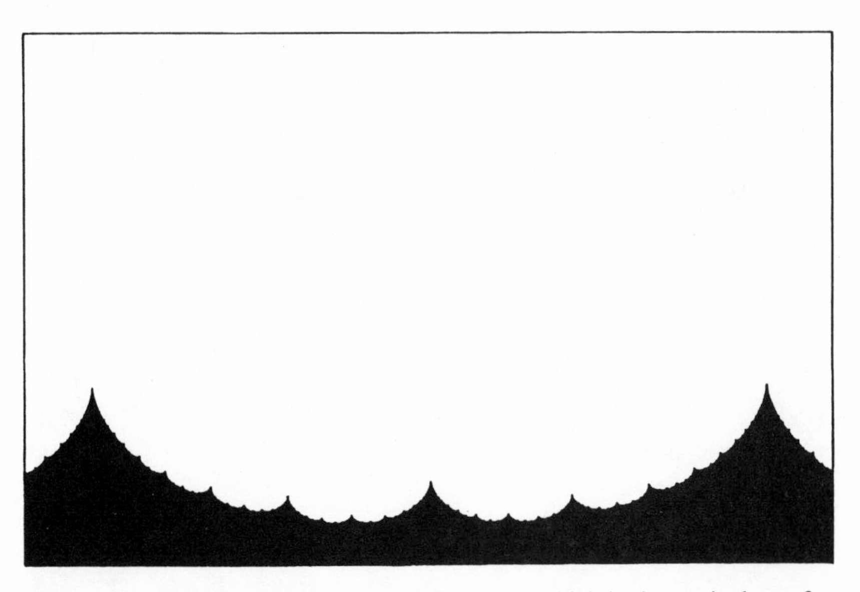

Figure 11.7 The limit set of a Teichmüller space which is the equivalent of a Mandelbrot set for Kleinian groups.

David Wright. Although it is not so complicated as the Mandelbrot set, there are still many unanswered questions about its shape. Careful study is leading to new insights about the way in which the limit-set pictures vary as they are distorted.

Mathematicians will often tell you that they work on their subject because it is beautiful. It is sometimes frustrating not to be able to explain what this means. Over the past few years, computer graphics have come to our aid. The kind of pictures I have been discussing are surely beautiful by any standards – there is debate as to whether they should be counted as a new art form. The mathematical ideas and relations that govern the pictures, telling us what to draw and what features to study are, to the mathematician, every bit as beautiful as the pictures themselves. Understanding the pictures may well give us new insights into the geometry and mechanisms of chaos: experience shows that the mathematical beauty of today usually translates into the useful tool of tomorrow.

12
Chaos, catastrophes and engineering

ALLAN MCROBIE AND MICHAEL THOMPSON

Applying chaos theory to engineering may seem odd. We expect machines and structures to operate with clockwork predictability. But engineers are now discovering that a sudden disaster may mean there's a fractal in the works.

In February 1974, the *Gaul*, a trawler from Hull, disappeared in heavy seas off the coast of Norway with the loss of 36 lives. The inquiry into the disaster found that there was nothing wrong with the ship's design: the stability of the vessel against capsize conformed with all required standards. One possible explanation put forward for the loss of the vessel was that it capsized as the result of some transient phenomenon when hit broadside by a short succession of abnormally large waves.

'Transient phenomenon' means that the vessel was moving in an irregular fashion: rocking, swaying, pitching, heaving and rolling. A trawler in a storm is obviously tossed around, so any capsize caused by waves hitting a vessel is a transient phenomenon. Strangely, though, the methods that engineers use to assess how stable a ship is assume that the ship is totally motionless. The methods simply require that the forces on the stationary vessel when tilted at various angles should be such as to try to right it. Various naval engineers at the inquiry pointed out that these tried and trusted methods of stability design might not always be safe. The motion of a ship and the way that it rocks change its stability. Engineers suggested that because of the

Figure 12.1 Transient motions are probably responsible for capsizing a vessel.

dynamic nature of capsize a ship's stability should be evaluated when it is moving as well as when it is still.

The word 'transient', however, has a stronger meaning than just motion: it is a mathematical term for unsteady motion – motion that has not yet settled down to a steady and regular pattern. The mathematics of transient motions gives equations that are more complicated than the equations of steady rocking, which are often complicated enough. To many engineers, assessing dynamic stability would mean analysing the steady rocking motions. Ships capsize, however, as a result of transient motions, so these should be considered as well.

Everything moves, even buildings. Many people in Britain first realized this in recent storms when office workers were evacuated from tower blocks shaken about by high winds.

Although the motion is usually imperceptible, engineers spend a lot of effort analysing in detail the way a building will move; the whole safety of the building and its occupants depends on an engineer checking that the building will not wobble in a high wind so much as to collapse.

At this moment, engineers around the world are busily analysing the way nuclear power stations rock during earthquakes, how offshore oil platforms behave when buffeted by waves, even how the next models of computer printers vibrate when working.

Most systems that confront the engineer are examples of 'nonlinear' dynamical systems. Nonlinear means that the parameters of the system do not vary proportionally. A graph representing this variation will not be a straight line. To simplify things, engineers often assume that such systems are linear. Often this is a good approximation. The force with which a material resists being stretched, for example, is roughly proportional to the amount that it is stretched. Stretch it too far, though, and the approximation is not so good. Typically, the material will stretch more easily, and eventually it may break. This is important: the linear approximation is usually worst when things are about to fail.

Most engineering systems also experience random forces. The forces that will act on the system are never known precisely beforehand and they will fluctuate throughout the life of the system.

You need only envisage the motion of several large waves breaking across the deck of a trawler to appreciate just how difficult the complete analysis of a real nonlinear, random, dynamic situation can be. One offal chute accidentally left open can easily invalidate an analysis. Faced with such complexity, the engineer simplifies the calculations. Ship designers assume that the vessel is still (the static approximation), but they retain the nonlinear features of the acting forces in their analysis. Designing modern high-rise buildings to withstand wind involves some comprehensive random dynamics but assumes that material properties of the building behave proportionally (the linear approximation).

Most engineering firms have computer packages that can take into account nonlinear and random dynamic effects. The real skill in engineering, however, is not in solving lots of complicated equations but in being able to identify the important principles and parameters in the face of the overabundance of detail and incompleteness of data. The major benefit of both the static and the linear approximation is not that they are mathematically easier to deal with but that such systems are well understood. Experienced engineers have a 'feel' for the way they behave. It is in similar terms of understanding that chaos theory has much to offer.

If we assume that the system is linear, we can represent the average forces acting on it by a parabolic potential well. This means that the system will obey laws of motion similar to those of a ball rolling in a trough shaped like a parabola. One implication of this shape is that the ball can never escape. The potential wells of real systems are not in general parabolics, however (see Figure 12.2). Most wells allow the possibility of escape – which corresponds to an engineering failure, be it a tower block

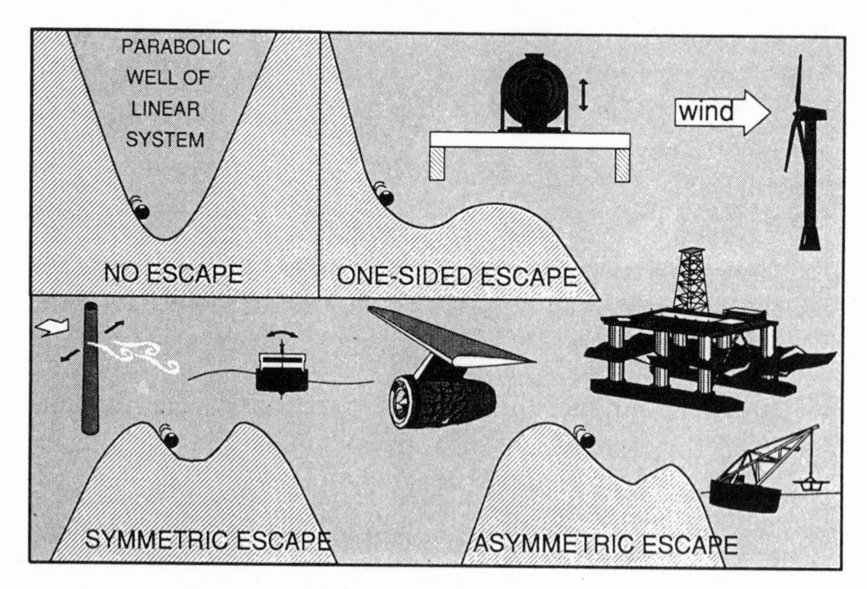

Figure 12.2 The stability of an engineering structure depends on the kind of 'potential well' it lives in. Escaping from the well corresponds to disaster.

collapsing or a ship capsizing. Consider the ship in Figure 12.2. Small rocking motions are like the ball rolling to and fro in the bottom of the well. If the oscillations become too large, then just as the ball may roll out of the well and escape to the left or the right, so the ship may capsize to starboard or port.

If you apply a periodic force to a linear system, the system will rock, or oscillate, erratically for a while – these are the transients – before settling down to a steady vibration with the same frequency as the applied force. This final motion does not depend on the system's initial conditions – its velocity and position when the force was first applied. There is only one long-term behaviour, called the steady-state solution.

If we include the nonlinearities of the system in the analysis, then more than one steady-state solution is produced, and the one on which the system settles will depend on its initial conditions. Such a solution need not have the same periodic time as the applied force, although it will usually be some multiple of it. Occasionally, however, when the transients decay, the system may not settle into any regular oscillation but may continue to vibrate erratically, never quite repeating any earlier manoeuvre and yet never escaping from the well. This is a chaotic solution, otherwise called a strange attractor. It can be represented graphically, so as to reveal its detailed form and inner structure, in a way that distinguishes chaos from the fuzziness arising from 'noisy' external forces.

Starting from different initial conditions, similar systems will have different transient behaviour but may eventually settle to the same chaotic solution. They will probably not be in step with each other but the nature of the long-term irregularity will be identical. They will have the same chaotic structure. Ian Stewart describes several examples of such chaotic solutions in Chapter 4. Such solutions are undoubtedly very interesting but, in terms of safety, the solution that the engineer is usually most concerned about is the 'attractor at infinity'. This is where the system can escape to. The transients of this attractor do not decay but grow without limit: the system escapes from the potential well and fails, collapses or capsizes.

The computer graphics shown in Plates 23 and 24 are examples of one way to represent how a nonlinear system responds to a periodic force. They are based on a technique called cell-to-cell mapping developed by Chieh Su Hsu at the University of California at Berkeley. The method is particularly useful for personal computers which can store a lot of information in the colours of the pixels forming the graphics screen. The blue regions, of either shade, in the centre-left picture of Plate 24 correspond to those initial conditions from which the system escaped out of the well; the yellow regions signify systems that converged to a steady-state solution of the same period as the applied force; and the red regions converged to a different solution at twice the period. Each steady-state solution is referred to as an attractor and is represented by only as many points as its period. The surrounding region that converges to that attractor is called the basin of that attractor. Plate 23 also shows an attractor with period eight, and its basin is coloured green. For the corresponding linear system, there would be one black pixel representing the unique steady-state and the rest of the screen would be yellow, because everything would converge to it. Not as pretty, and not as realistic.

Most of the basin boundaries in Plates 23 and 24 are fractals. This means that no matter how much you magnify the boundary regions a structure made up of infinitely finer layers is revealed.

By examining many such pictures for different parameters – frequency, amplitude of forcing, for example – we can obtain an overview of how the system behaves. Figure 12.3 shows the amplitudes of the main attractors plotted against frequency for three similar systems. In the linear system, there is only one response at any frequency, and the sharp peak of the graph where the amplitude of the response becomes large is known as resonance. The nonlinear system, corresponding to a one-sided escape well similar to the one shown in Figure 12.2, has a range of frequencies where two stable period-one solutions coexist, one with a small and one with a large amplitude. The peak of the graph has bent over to the left, and near the top of the curve there are period-doubling cascades similar to the Feigenbaum

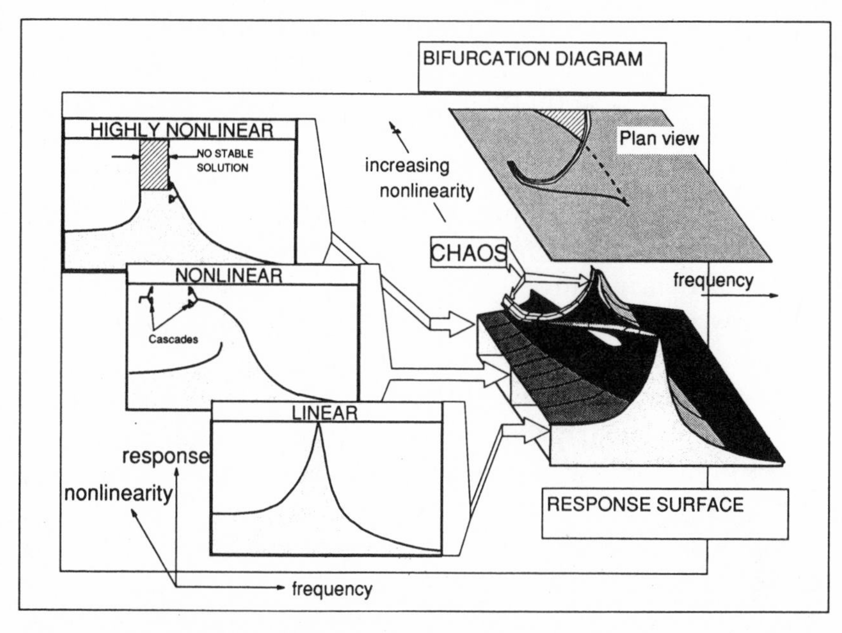

Figure 12.3 The steady response of a system to a periodic force shows a thin chaotic rim as the system becomes more nonlinear.

cascades encountered in the simple iterative equations discussed in Chapter 3. At the end of each cascade, there is a narrow band of frequencies where a chaotic solution exists. Adjust the frequency a little more and the chaotic solution vanishes in an event called a 'blue sky catastrophe'.

With gradual changes in parameter, attractors generally evolve smoothly, but at certain critical points, called bifurcations, the attractor may split into different attractors or may simply disappear. Such an event is a catastrophe in the mathematical sense, introduced by the French mathematician René Thom, at the Institute for Advanced Scientific Studies in Bures-sur-Yvette, France, who used powerful theorems to show how such catastrophes could be classified into a small number of elementary forms. It is a catastrophe in the everyday sense only if it means that a system close to the attractor can suddenly head for the attractor at infinity. Some catastrophes are relatively safe in that they need not imply collapse, merely a jump from one stable attractor to another.

Figure 12.4a shows a good example. It may be more dangerous to slow down a rotating machine than to start it up, because start-up may pass through a safe catastrophe, whereas wind-down might lead to an unsafe blue sky catastrophe. In this case, these are not yet hard and fast rules, and studies of these jumps are giving some interesting results. Even the first jump may not always be safe, it seems. Figure 12.4b shows the response of a ship experiencing waves of gradually increasing height. It reveals an early safe catastrophe before the dangerous blue sky event where capsize is inevitable.

In engineering, the parameters rarely vary slowly enough and smoothly enough for the system to adhere rigorously to the paths of attractors. The applied force inevitably contains some noise. Any real system attempting to follow the steady-state solution paths (as in Figure 12.4) may escape long before reaching the period-doubling cascades. We can understand the reasons for this by looking not just at the attractors, but also at the basins, their boundaries and how they evolve.

Around any attractor there is an area that converges to that attractor. This region may be so thin or wispy that small dis-

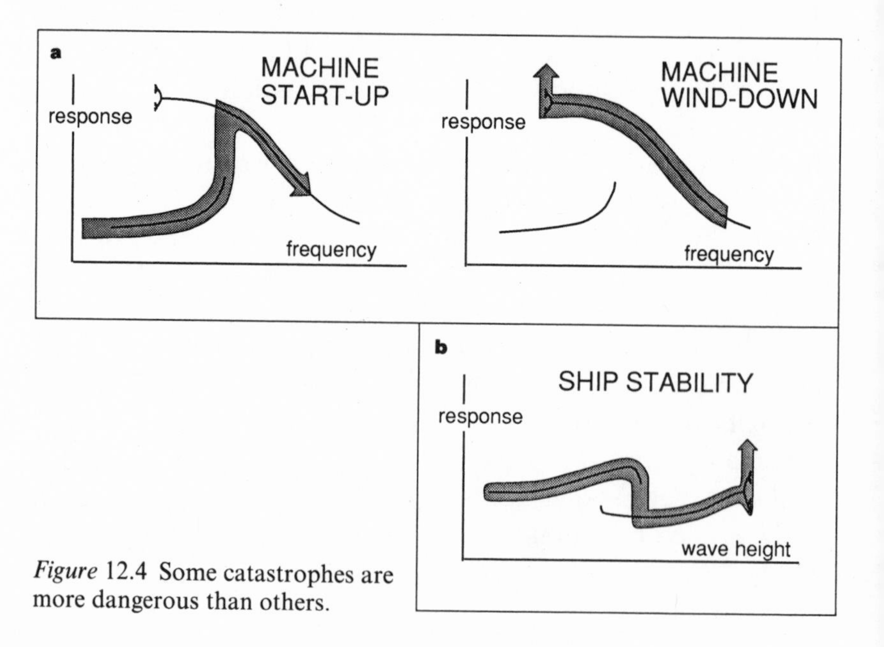

Figure 12.4 Some catastrophes are more dangerous than others.

turbances can knock the system out of the basin to converge somewhere else. To be safe, any engineering system must have its attractor at a prudent distance from its basin boundary (see Plate 25). Structural engineers refer to failure as 'exceedance of the ultimate limit state'. Once a structure passes this condition, it will collapse. In the dynamical case, the ultimate limit state is the outer basin boundary, where the blue meets the yellow and red. Beyond that boundary, the system must escape. The ultimate limit state is therefore often a fractal. Not a lot of engineers are aware of this yet.

We have used various 'integrity measures' to quantify the margin of safety between an attractor and the basin boundary (see Plate 25). In studies of how ships capsize, we have observed these measures to decrease very dramatically at a point on the solution path at which the realistic, noisy system is liable to escape. We refer to the phenomenon as the 'Dover cliff' effect. It occurs when the fractal 'fingers' of the basin boundary suddenly intrude deep into the centre of the basin. The effect happens irrespective of the exact equation describing the potential well. Provided the well looks rather like the one-sided escape well of Figure 12.2 with one trough and one hump, the effect appears to persist.

Most mathematical approaches to nonlinear dynamical systems focus on finding steady-state solutions and resonant motions of large amplitude. A trawler may capsize, however, as a result of a few abnormally large waves during which a steady rocking motion has not had a chance to develop. In terms of the ultimate safety, the emphasis of the analysis thus shifts. The steady-state solutions are almost irrelevant; the transient behaviour is where the action is. Did the ship capsize or not? Engineers can examine the computer graphics, not to look for what or where the attractors are, but to see what the transients did. How many transients led to failure? Because many engineering failures happen within the first few transients, particularly in ship stability, methods even more straightforward than cell-to-cell mapping can be used.

Using simple computer programs, engineers can thus readily

determine when any specific design is liable to fail. But safe engineering should not just rely on a computer printout. It must involve understanding how a design will behave and how and why it can fail. This is where chaos theory and the study of the fractals in all their exquisite mathematical intricacy is unmatched by earlier methods.

The fractal basin boundary is just one part of a geometrical structure called a homoclinic tangle (see 'Homoclinic tangles, horseshoe maps and holistic engineering', below) and the inner workings of these tangles can explain a lot of the features of nonlinear dynamic behaviour.

Homoclinic tangles, horseshoe maps and holistic engineering

Around the turn of the century, the great French mathematician Henri Poincaré first glimpsed the structure of a homoclinic tangle. He was, apparently, horrified. Homoclinic tangles occur in celestial mechanics, where Poincaré first met them on Saturn's rings. They also occur in engineering. They are formed by two fractal structures overlapping, the 'inset' and 'outset' of an unstable solution called a saddle.

In a simple one-sided escape system (Fig. 12.5), the inset corresponds to the global basin boundary. All points 'inside' the inset (normally yellow — white here) converge to some attractor and all points 'outside' it (normally blue — shaded here) escape to the attractor at infinity. The fractal nature of the curves is such that under any magnification a set of nested stripes-within-stripes can be observed. This striation is strongly related to the Cantor set, a fractal that is simple to construct but which has many unusual properties. The Cantor set was discovered in 1875 by Henry Smith and later used by the German mathematician Georg Cantor. Unlike the famous Mandelbrot set, which is an infinitely jagged fractal, the inset and outset fractals are smooth curves. Sensitive dependence on initial conditions results from this infinite layering.

Engineering systems dissipate

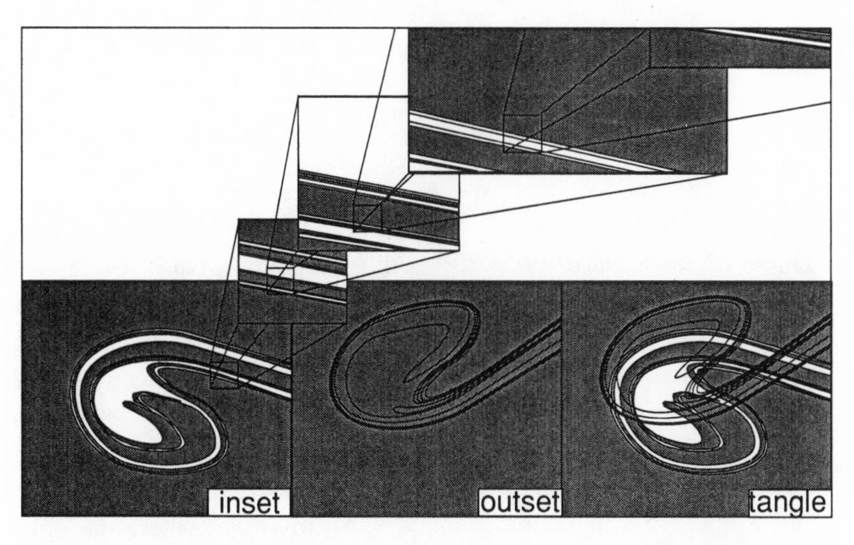

Figure 12.5 The complicated homoclinic tangle seen here is one of the complications of nonlinear dynamics. Its elaborate fractal structure reveals how real engineering systems work and sometimes fail. The white area is safe; the shaded area means failure.

energy. One result of this is that although the outset and inset may appear to be somewhat similar in shape, the inset has infinite area 'inside' while the outset has zero area inside. The outset, therefore, has the peculiar property of being an infinitely long, smooth, connected perimeter that bites out all of its own inside. It may not look like that in Figure 12.5, but only the first few bites have been drawn. On occasion, the inset can do the same thing. This is the phenomenon of basin erosion. The fractal inset (the boundary) intrudes deeper and deeper into its interior basin, until at the blue sky catastrophe (between chaos and the zone of no stable solution), the boundary succeeds in biting out all of its own inside, too. There is no white left and nothing is safe.

Before chaos theory, nonlinear dynamic problems were mostly tackled by perturbation theory. It usually involved a lot of algebra, long equations and lots of sines and cosines. Chaos theory uses a very different language, the language of topology.

This is a language engineers do not usually learn. Using this very rigorous and pure mathematics, you can learn things about dynamics that do not depend on the exact equations but rather on the 'shape' of the graphs and the 'form' of the equations. This is the concept of 'universality'. The power of chaos theory is embodied in this concept. What chaos theory implies is: 'If the equations have these properties then the dynamic response will have these features.' This is very important in engineering where you never know the exact equations and values.

Engineers may ask: 'How can something so mathematically precise and detailed as chaos be of any help when we never know our equations exactly?' The answer lies in 'universality'. While ignoring many of the details of the system, chaos theory can make detailed predictions about how it can and cannot behave.

What is more, chaos theory applies to the cumbersome, mathematically messy but realistic equations of engineering. Do not worry about the fine details of your system, just look at the shapes.

One example of this approach is called 'symbolic dynamics'. It relies heavily on the work of the American mathematician Stephen Smale and his construction called the 'horseshoe map'. It refers to regions of the graphics, the homoclinic tangles, in which some number of forcing cycles later map back across themselves in the shape of a horseshoe. The fascinating abstraction of Smale was to show that much of the important dynamics in those regions is equivalent to all the possible ways of writing down infinite sequences of two symbols (0 and 1, say) on either side of a decimal point, for example: strings of the form ... 01011101011010·0010011110100 ... At any point where the inset crosses the outset, there is a horseshoe, and the symbol string above represents one state of the system. Move the decimal point one place to the right and you get the state of the system one forcing cycle later. This is called a Bernoulli shift. Dynamic analysis becomes an exercise in simple binary logic, writing down sequences of 0s and 1s and moving decimal points. This is very different from the pages of sines and cosines that engineers are used to.

Using such holistic techniques, we are beginning to understand many of the features of the response graphs such as Figure 12.3. Moreover, such methods provide a way to analyse, quantify and understand the mechanism of basin erosion, with all its implications for engineering safety.

The archetypal manifestation of chaos is the strange attractor and these, it seems, do not occur very often in engineering. You can find strange attractors in impacting systems where the smooth dynamics is interrupted by hitting against some nearby body or stop. They happen where the forces arise internally rather than from simple external application. Very often though, strange attractors exist only over extremely narrow ranges of frequency and require the sort of ideal conditions that few real engineering systems can provide. Chaos, however, exerts a far stronger influence than this would suggest. Not only does a strange attractor often herald a dangerous catastrophe, but there exists other more important and more common chaotic solutions. These solutions attract only in some directions and repel in others. They occur throughout homoclinic tangles.

The most far-reaching benefit of applying chaos theory to engineering is that it is helping us to understand nonlinear dynamic behaviour in real situations. When nonlinearities are included in engineering dynamics, many of the familiar landmarks of linear analysis are left behind. Good engineering demands that engineers have an intuitive feel for how their designs behave, rather than just relying on a set of numbers. The study of the shapes of the underlying fractal structures and the rules that govern their evolution can provide just such a deeper understanding.

Further reading

J. M. T. THOMPSON and H. B. STEWART, *Nonlinear Dynamics and Chaos – Geometrical Methods for Engineers and Scientists*, Wiley 1986.

J. M. T. THOMPSON, R. C. T. RAINEY and M. S. SOLIMAN, *Philosophical Transactions of the Royal Society*, London, A332, 1990, p. 49.

J. M. T. THOMPSON, *Proceedings of the Royal Society*, London, A421, 1989, p. 195.

13
Chaos on the circuit board

JIM LESURF

No electronic system is perfect. Engineers are using chaos theory to help them to understand the wayward behaviour of electrical circuits.

Engineers prefer electronic circuits that behave predictably. When designing digital logic, they tend to choose circuits that perform well-defined functions such as AND, OR, or NOT. A careful search through the catalogues of most chip manufacturers will fail to find any PROBABLY, SOMETIMES, or WHYNOT functions. This is because it is difficult or impossible to predict just what a collection of such circuits would do when switched on.

Similarly, if we buy a watch, we expect it to keep 'good time'. The hands should move or the displayed numbers should change at regular intervals. A clock whose hands moved unpredictably faster or slower, perhaps even sometimes going backwards, would not be very useful. As a result, most books on electronics concentrate on the things that 'work well', in other words, reliably. Despite this, it is surprisingly easy to devise electronic systems that behave 'badly', or chaotically. Now electronics engineers have come up with some neat ways of dealing with chaos in electronic circuits. Occasionally, they even use it constructively.

Transistors operate by responding to a change in the signal, a voltage or current, altering their output signal. A graph showing how the output changes as we alter the input gives us an idea of how the transistor behaves. A graph of this type is called the characteristic curve of the device. The 'gain' of the transistor, or any similar device, can then be calculated from the slope of the line that shows how much the output will change if we alter the input.

A common feature of all electronic systems, from hi-fi amplifiers to powerful computers, is that they exploit devices that show gain. Most modern electronic systems use various types of transistor. They also use devices such as valves and various types of semiconductor diode. We can describe how these devices behave in terms of how they respond to changes in an applied input voltage or current.

When a hi-fi company develops an amplifier, their main concern is that its output should simply be a larger version of the input. Ideally, an amplifier's behaviour could then be specified by just one number, its overall gain, G. If we plotted the output from such an amplifier against the input, we would discover that the result was a straight line whose slope was equal to G. Such an amplifier would be described as linear.

Alas, any real electronic device is nonlinear. Its output will not simply be proportional to the input, and so its characteristic plot is not a straight line. No real amplifier can ever be absolutely linear, because its behaviour will depend upon the detailed properties of all the devices from which it has been assembled. Nonlinearity is a problem in many situations. In hi-fi, for example, nonlinearity causes the output from an amplifier to become distorted so that what we hear is no longer precisely what was intended.

We can attempt to deal with nonlinearity in two ways. First, we can try to find devices that are inherently more linear – in other words, their characteristic plot looks as much like a straight line as possible. Secondly, we can apply an ingenious technique called feedback.

The principle behind feedback is quite simple: we arrange for the amplifier to be able to compare its output with its input.

Take, for example, an amplifier that is nonlinear, but which has a gain roughly equal to G for most input signals. We arrange for part of the amplifier's output to be returned or fed back to its input for comparison with the original. Because we expect the output to be larger than the original input we must reduce the returned signal by an amount $1/G$ in order to make a fair comparison. This is done by passing the signal through a system designed to attenuate that signal by an amount, α (which, it is hoped, equals $1/G$).

The amplifier can compare the fed-back signal with the original, and any difference between them can be used to correct the output, producing a better result. This process can never completely banish nonlinearity because the amplifier must always see some small 'error' between the desired and actual output in order to let the circuit know a correction is needed. Despite this, feedback can often reduce signal distortions to a thousandth or less of their uncorrected level.

Feedback can be tremendously helpful in reducing nonlinearities. But it must be applied with care, because adding feedback to a nonlinear circuit with gain is a recipe for chaos. In reality, any electronic device or circuit takes a finite time to respond to a change at its input. This gives us a problem because we have assumed that the original and the 'fed-back' comparison signal appear simultaneously, which is impossible. Any comparison signal taken from an amplifier's output must take some time, however small, to get back to the input. This may lead to an amplifier that does not always do what we expect.

In some cases, these delays can be very useful. Many oscillators, including those in digital watches, make use of the effects of feeding back delayed signals. To see how this works, imagine what happens if we inject a brief pulse into the combination of an amplifier and a feedback network. For example, consider an amplifier of gain, G, which takes a time, t, to react to an input, being used with a feedback arrangement which takes a time, τ, to return a signal while reducing it by an amount, α.

We start off by injecting a short pulse into the amplifier's input. A time, t, later the output produces a pulse which is G times larger. Part of this travels back to the input. As a result, a time $(t+\tau)$ after the initial pulse, another fed-back 'echoed' pulse, whose size is G multiplied by α bigger than the original, appears at the amplifier's input. This, in turn, passes through the amplifier and produces another output pulse after a further delay. This is fed back, and so on. The result is a chain of re-echoed pulses reverberating around the amplifier and its feedback arrangement.

The behaviour of the system now depends upon the size of the gain, G, multiplied by the feedback, α. If we have ensured that the reduction, α, exactly equals $1/G$, then G times α will be 1. This means that each echoed pulse will be identical in size with the initial one. The result is a chain of identical pulses occurring at regular intervals, $T=(t+\tau)$. We have built a clock or oscillator that produces an output which repeats itself over and over again in a regular way with a fixed period, T.

In fact, when comparing the original and fed-back signals we are interested in the difference between them – we want to subtract one from the other. So, we generally arrange for G or α to be negative (a positive input produces a negative echo and vice versa). This means that the repeating pulses are alternately inverted. As a result, the output then repeats itself with a period $2T$.

Unfortunately, we may well find in practice that the size of feedback reduction, α, is slightly bigger or smaller than the amplifier gain, G. The behaviour of the circuit then depends upon the size of α multiplied by G. If this is less than 1, each pulse is smaller than the last and the echoes fade away rapidly. If it is greater than 1, the echoes grow steadily.

If we apply a constant input, x, the output will initially be equal to x times G. After a time, this produces an echo and the output becomes $xG(1+\alpha G)$, which in turn produces an echo, and so on. Provided that α times G is small, this series will add up to a finite number and the amplifier output will settle on a particular value. If α times G is too large, however, the series of echoes may try to add up to an infinite result.

In practice, the output cannot go on growing forever. The amplifier will be powered by batteries or a mains supply which provide it with some maximum voltage and current. This means that the amplifier cannot produce an output greater than this maximum level. At some point, a series of growing pulses will reach this maximum and stop growing. This, in itself, is a sort of nonlinearity because we find that, from now on, the output produced by a given input must be less than G times bigger.

You can analyse the behaviour of the amplifier and its feedback in the same way as other chaotic systems. An input signal, x_0, leads to an output, y_0, which in turn produces a new input, x_1, which leads to a new output, y_1, and so on. Each output, y_n, produces a new output, y_{n+1}. We can describe this process in terms of what is called a logistic mapping: $y_i \rightarrow y_{i+1}$. Franco Vivaldi described the logistic map in detail in Chapter 3.

Here, we can think of the arrow $\rightarrow$ as meaning 'produces'. For a linear amplifier and feedback system we can say that an output, y_i, will produce an output echo $y_{i+1} = \alpha G y_i$. In this case, the appropriate logistical mapping would be: $y \rightarrow \alpha G y$. To see what happens when we apply a particular initial input, we can just repeatedly 'update' the output by calculating the change produced by this mapping over and over again.

Because the output from any practical amplifier cannot move outside some finite range, we can take this into account by replacing the constant value of αG with some other value that itself depends upon the size of y. Variation in this value with y can then represent the nonlinearity of the amplifier and its feedback arrangement. Although the details of these nonlinearities will differ from one system to another, we can expect that there will always be a tendency for this value to fall when y becomes very large. Otherwise the amplifier might be capable of producing infinite output.

One of the simplest mappings that can produce chaotic results is: $y \rightarrow ky(1-y)$. Comparing this with the above we can see that this is the equivalent to building an amplifier whose echo gain is $k(1-y)$, in other words, it falls as the output level, y, increases. If we repeatedly use this simple expression to produce new values for y, we find that the result depends critically upon the value chosen for k. Provided that k is between zero and three, y tends to reach some fixed value. When k is between three and four, y hops around repeatedly between a set of values, never settling to any one in particular. When k is greater than four, y jumps about in an apparently patternless, chaotic manner, never visiting any value more than once.

For most purposes, engineers concentrate on making amplifiers that have a fairly constant gain over some particular range of output values. They then try to arrange that the input given to the amplifier is always kept within limits that ensure that the output never has to try and reach unattainable levels. We can, however, deliberately produce circuits that have quite severe nonlinearities. These can sometimes be used to produce chaotic effects (see 'Every chaotic circuit is an individual', below).

Every chaotic circuit is an individual

Figure 13.1 shows the circuit diagram of a fairly simple transistor amplifier. Amplifiers of this type are used in many pieces of electronic equipment. They produce quite acceptable results, provided you keep the input signal small and the feedback arrangement fairly simple. The design, however, is inherently a 'poor' one and will show various sorts of nonlinearity if the input signal level is allowed to cover too large a range. Here the amplifier is shown with a fairly complex feedback arrangement of inductors ($L1$ to 3) and capacitors ($C1$ to 3, Cs) which make chaotic behaviour a distinct possibility.

A detailed analysis of this amplifier would be quite difficult. Despite this, we can fairly easily explain the two main effects that make the amplifier nonlinear. One of these can be seen by looking

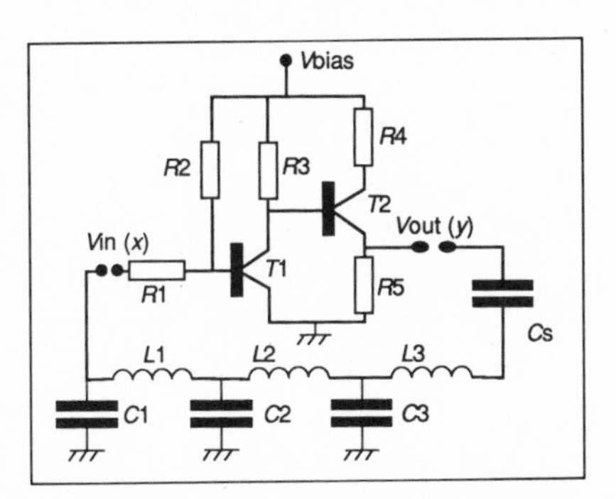

Figure 13.1 Diagram of a nonlinear amplifier. The output voltage first rises, then falls as the input increases. Component values: $R1 = 2700$ ohms, $R3 = R4 = R5 = 270$ ohms (all ± 5 per cent); $R2$ was a variable resistor to cover a range of 10 to 100 kilo-ohms; $T1 = \text{BC109}$, $T2 = \text{BC477}$; $Cs = C1 = C2 = C3 = 0{\cdot}01$ microfarads (all ± 20 per cent); $L1 = 5{\cdot}2$ millihenries, $L2 = 3{\cdot}5$ millihenries, $L3 = 4{\cdot}9$ millihenries (all ± 10 per cent).

at its voltage characteristic curve. This shows that, at low voltages, as the input voltage, x, is increased, the output voltage, y, also rises rapidly. This is because the increasing input causes extra current to flow through both transistors, boosting the output.

Eventually, the output transistor, $T2$, 'saturates' and it cannot produce any extra current. Any further increase in the input makes the first transistor, $T1$, 'drag down' the output. Hence, above some critical level, any rise in the input causes the output to fall. This process finally stops when $T1$ also saturates.

The other main nonlinearity comes from the fact that an ordinary transistor has an input that shares some of the character of a diode. Current can flow through the transistor more easily in one direction than the other. As a result the circuit tends to 'rectify' the input signal it sees.

Usually in electronics it is easy to decide whether the choice of components (transistors, resistors, and so on) is critical or not. For circuits intended to be chaotic the situation becomes rather odd. This is because no two components will be identical – even if they are marked as being the same. A pair of resistors may both be marked '270 ohms', but if we measure them precisely enough we will always find that their values differ. They may be similar to one part in a million or a billion, but never identical.

Most electronic components are normally specified with an accuracy of only a few per cent. Usually, this does not matter but with chaotic systems it means that no two 'identical' circuits will ever do precisely the same thing. I built a test 'chaotic oscillator' using components listed above. We found that varying the applied bias voltage over a range of 5 to 20 volts, or slowly altering the value of $R2$, made the circuit pass through alternating regions of chaos and regular, periodic oscillations.

Another circuit, built using similar components, would behave in a similar way, but the exact details of its behaviour would be different. A circuit made using very different components would show its own pattern of behaviour. When it comes to chaos, every component, every circuit, is an individual.

What happens when we feed back gain is important for designers, who generally want linear amplifiers. It is also of critical importance to anyone who wants to design oscillators – circuits designed to generate regular, periodic output signals.

Electronic engineers employ two approaches when making oscillators. The most common is to combine an amplifier with a selective feedback arrangement. We can see that, in principle, any waveform that repeats itself after the right period will be re-echoed. Hence, if we want to ensure that an oscillator produces a specific sort of output – be it a series of pulses or a sine wave – we have to make the system 'prefer' the waveshape we want. We can do this by building a feedback circuit that passes only the waveshape that we require.

This approach works well provided that the amplifier is fairly linear. If it is very nonlinear, however, each successive pass through the amplifier seriously changes or distorts the shape of the repeating wave. What happens then depends upon the result of a fight between the amplifier and the feedback. If the feedback arrangement can reject the unwanted distortions and still see enough of the original shape to return one more to the amplifier, then the system continues to re-echo the same wave. If, however, the feedback circuit is unsuccessful, we find that each new echo may differ noticeably from the last.

To add insult to injury, we can use an unselective feedback arrangement that delays different types of signal by differing amounts. Then both the nonlinear amplifier and the feedback will tend to scramble the echoed signals. If the system is 'bad' enough, the result may be a chaotic oscillation. The arrangement shown above ('Every chaotic circuit is an individual') works in this way.

At very high frequencies, it becomes almost impossible to make conventional electronic amplifiers. As a consequence, microwave and millimetre-wave engineers tend to use the properties of individual devices instead of collections of transistors. For frequencies above about 50 gigahertz, we can use solid-state devices – for example, Gunn diodes – which show negative resistance. Just as an ordinary, positive resistance tends to dissipate electrical oscillations, so a negative resistance tends to generate oscillations.

Millimetre-wave negative resistance diodes usually make use of the properties of fairly esoteric materials and complex semiconductor structures. They are also quite expensive and normally operate only at high frequencies. Fortunately, it is fairly easy to build simple transistor circuits that behave as negative resistances at audible frequencies (see 'Negative resistance generates chaos', below).

Most houses contain quite a few very simple nonlinear electronic devices. Ordinary incandescent light bulbs have a resistance that increases with the applied voltage. Even more interesting are fluorescent lights and neon indicators, which have a resistance that falls when the voltage is increased. As a consequence, neon indicators can be used to make a type of negative resistance oscillator, and fluorescent tubes have to be designed so as not to interfere with radio and TV reception.

Usually, engineers do not want chaotic circuits – although they sometimes find they have built one by accident. The performance of a microwave oscillator depends on its basic frequency of oscillation and the amount of power it produces at that frequency. For most purposes, we require oscillators that produce a 'pure' sine wave at just one frequency.

In practice, all real oscillators suffer from two main defects. First, they tend to produce 'extra' oscillations at various frequencies. Secondly, their main, wanted output tends to fluctuate unpredictably in both power and frequency. Until recently, engineers have regarded these problems as separate. They considered unwanted oscillations as the product of distortions resulting from nonlinearity. Random fluctuations in the wanted output were considered as being due to random noise. This noise arises because of random movements of the electrons in any conductor or semiconductor.

Viewed with a chaotic eye, however, the picture looks subtly different. Given a nonlinear device that can amplify signals (a diode in this case) and a complex feedback arrangement, we have to accept the risk that the output of the circuit may not

Negative resistance generates chaos

Electronic engineers use negative resistance devices to make microwave and millimetre-wave oscillators. Carefully designed oscillators can produce sinewaves at frequencies up to about 100 gigahertz. As with amplifiers, you can produce 'bad' devices and feedback circuits that behave chaotically.

These devices are essentially diodes, so they cannot be used in exactly the same way as an amplifier, because their output does not appear in a different place from the input. Instead, engineers use a feedback arrangement that is sensitive to the change in current resulting from altering the input voltage. The feedback circuit is designed to respond to a current change by producing a change in the voltage after a short time delay.

Any change in the input voltage produces an alteration in the current, once the device has had time to respond. This change in current passes through the feedback arrangement and results in a further voltage change after another delay. If the device behaves as a negative resistance, so that the current falls when the voltage rises, this combination may lead to variations that persist in much the same way as the 'echoes' in a conventional combination of an amplifier with a suitable feedback circuit.

Figure 13.2 shows a transistor circuit that you can use to mimic

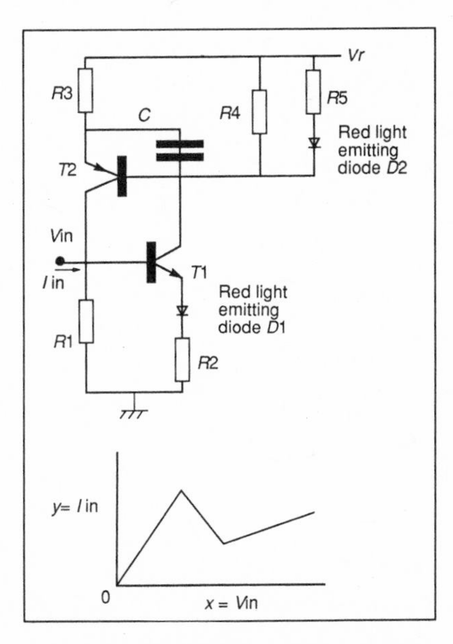

Figure 13.2 Diagram of an artificial 'negative resistance' device. Typical component values: $R1 = R2 = R4 = 270$ ohms, $R3 = 100$ ohms (all ± 5 per cent); $C1 = 0{\cdot}01$ microfarads (± 20 per cent); $T1 =$ BC109, $T2 =$ BC477, two red light-emitting devices; Vr = between 8 and 20 volts. This circuit shows negative resistance for Vin between about 3 and 4 volts.

how a negative resistance device behaves. When the applied input voltage, Vin, is small, no current will pass through either of the two transistors. Under these circumstances the input current is simply given by Ohm's law, $I\text{in} = V\text{in}/R1$.

When Vin rises to a 'peak' level the transistors will begin to conduct. (In the circuit shown, Vin will need to exceed about 3 volts to reach the 'turn-on' voltage of the transistors and the red light-emitting diode, $D1$.) Now any increase in Vin will generate 'extra' current from $T2$, which will pass through $R1$. If $R2$ equals $R4$ and $R3$ is less than or equal to $R1$, this produces more current than is required to increase Vin, and so Iin falls. As Vin is increased, we find that at a 'valley' level the second light-emitting diode, $D2$, will conduct and Iin begins to increase with Vin once again.

A small capacitor, $C1$, can be used to mimic the time taken for a real device to respond to a change in the applied voltage. It is then possible to experiment with the effects of supplying a range of voltages to the input through various types of feedback arrangement. Under most circumstances, the 'negative resistance' device and feedback arrangement either does nothing or oscillates in a periodic manner. You can, however, use the system to generate chaotic output. In particular, you can connect several of these devices together in a network in an attempt to combine their outputs. Such a system can produce very complex patterns of behaviour. The component values shown will give a system that behaves as a negative device.

---

simply be an ideal, periodic signal. What is more, the apparently random movements of electrons can be seen as the result of millions of nonlinear interactions between all the electrons and the atoms they move between.

Although the subject is, for practical engineers, a fairly new one it begins to look as if chaos may help us to understand the behaviour of imperfect electronic systems. That is our main interest in the millimetre-wave group at the University of St Andrews. We are using chaos as a way of trying to understand how to make better, less chaotic oscillators. For this, studying chaos means 'know thy enemy'. It is very difficult to observe the detailed behaviour of an oscillator operating at 100 gigahertz because everything happens so quickly. The 'artificial' negative resistance shown above ('Negative resistance generates chaos') can be used to 'mimic' the millimetre-wave oscillators, but at frequencies below a few kilohertz.

This approach will allow us to develop new systems that avoid problems and behave in a more predictable, unchaotic manner. It may also allow us to invent entirely new systems that make use of enhanced forms of chaos to good effect. For engineers, the problem then will be: who wants chaos?

Well, sometimes engineers want to produce random, noise-like signals. Military communication systems, for example, sometimes transmit radio signals designed to 'hide' in the background noise. These cannot actually be random, otherwise they would not convey any information, but they should imitate real noise as closely as possible to avoid being noticed by eavesdroppers. The kinds of signals produced by 'chaotic oscillators' may be ideal for this.

14
Chaos on the trading floor

ROBERT SAVIT

Economists and speculators would like to be able to predict the ups and downs of the financial and commodities markets. Could chaos theory help?

It is a remarkable experience to visit the trading floor of a financial or commodities exchange. During an active period of trading, pandemonium reigns. Imagine the scene: hundreds of people waving their arms and shouting at the tops of their lungs trying to make the right transaction at exactly the right time, while trying to monitor the behaviour of their fellow traders and assimilate the new information assaulting them from every direction.

But underpinning this mêlée, there are some specific rules and motivations common to all the players. The primary goal is, of course, to make money, and to do it in the least risky way. Is there any rational description of this capitalist brawl, and of the capitalist economies in general, of which these markets are a reflection and a seminal part?

One important feature that all financial markets, and the larger economies, share is that they are systems with many feedback and self-regulatory mechanisms. We know, for example, that if the price of an item rises too high, demand for the item will decrease and the price will drop. This is the simplest of many such self-regulatory mechanisms in economics and finance. But the existence of these kinds of inherent mechanisms has profound and surprising implications for the ways in which markets, prices and economies could behave.

To see why this is so, consider a simple model of a system with self-regulating feedback. Suppose we have an extremely simple market with just one commodity, say gold, for sale. The price of an ounce of gold during a particular week t is $p(t)$. Suppose also that the gold dealers are rather greedy, and try to raise the price by a factor, A (bigger than 1), each week. Then the price during week $t+1$ will be given by the simple equation, $p(t+1)=Ap(t)$. But the consumers are also sensitive to price hikes, and as the price goes up, there are fewer buyers. We can encode this effect, qualitatively, in a simple way by subtracting from the right-hand side of the above equation some number that gets larger as p gets larger. If we subtract something of the form $Bp(t)$, then we have not really altered the basic form of the equation. That is, we now have:

$$p(t+1)=Ap(t)-Bp(t)=(A-B)p(t)$$

If $A-B$ is greater than 1, this equation will still lead to continued growth, so that the negative feedback (the reduced demand by consumers for the product which is too expensive) will not be sufficient to stem inflation. On the other hand, if $A-B$ is less than 1, the price will eventually go to zero: we have a depression (at least in the gold market).

Clearly, a model of a market in which the only two choices are indefinitely rising prices or the absolute collapse of the market is not very satisfactory. To try to remedy this, we can consider the next simplest thing that will still allow us to mimic the effects of consumer resistance to high prices. Rather than subtracting a term from the right-hand side of the equation that is linear in $p(t)$, we can subtract a term with a higher power of $p(t)$ in it. The simplest choice is $Ap^2(t)$, so that our equation becomes:

$$p(t+1)=Ap(t)-Ap^2(t).$$

This equation is the famous logistic map that previous chapters have described (see Figure 14.1). Despite its simplicity, the time series $p(t)$ generated by this map can have a staggering variety of behaviour depending on the value of A. It turns out that if A is between 0 and 3, then after a long enough time $p(t)$ will become constant. For somewhat larger values of A, the behaviour of $p(t)$ is described as a 'limit cycle'; $p(t)$ bounces

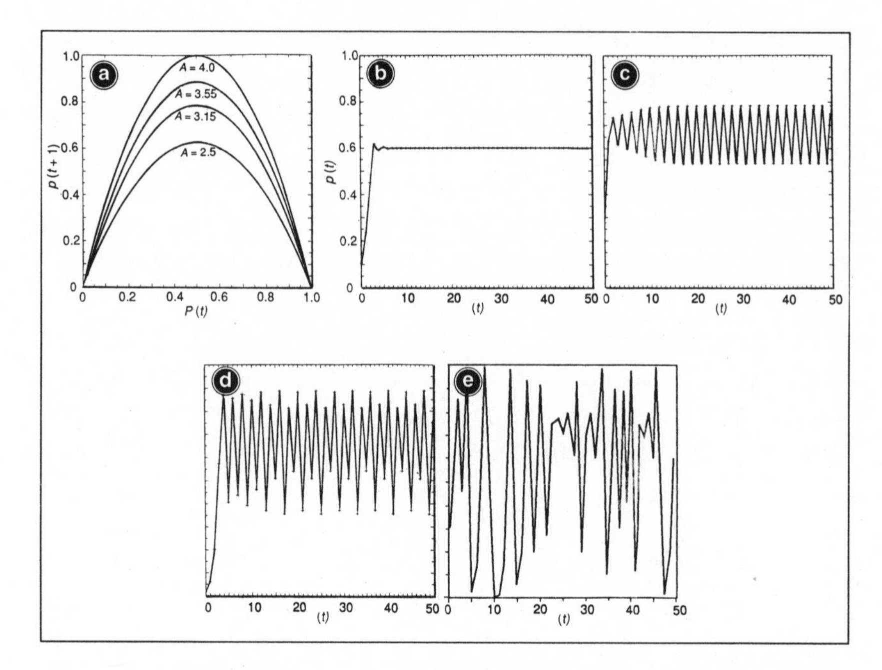

Figure 14.1 The logistic map in equation: $p(t+1)=Ap(t)-Ap^2(t)$ is just the simple parabola, familiar from high-school mathematics, shown in **a**. As we increase A, the parabola gets steeper, but its form does not change very much. Nevertheless, the series, $p(t)$, generated by the logistic map has dramatically different behaviour depending on the precise value of A. In **b** ($A=2{\cdot}5$), we see that $p(t)$ eventually settles down to a value of $p(t)=2{\cdot}6$. For a somewhat larger value of A ($A=3{\cdot}15$), $p(t)$ bounces indefinitely between two values as shown in **c**. This is the simplest example of a limit cycle. In this case, it is called a 2-cycle because the series bounces between two values.

For still larger values of A, $p(t)$ oscillates among more values. For $3<A<3{\cdot}5699$, $p(t)$ will oscillate between some number of values which is a multiple of 2. For example, in **d**, we show $p(t)$ for $A=3{\cdot}55$, where $p(t)$ oscillates indefinitely among eight values (an 8-cycle). For still larger values of A, the series $p(t)$ takes on many different kinds of unusual behaviour. One particularly interesting case is $A=4$, shown in **e**, for which the series $p(t)$ is chaotic. In this Figure, $p(t)$ never repeats itself, unlike the behaviour in **b** to **e**. This behaviour looks much more 'random' than the other graphs, and could easily be mistaken for a random sequence.

among several values periodically (the precise number of values depends on the value of A). Finally, for still larger values of A the behaviour of $p(t)$ becomes aperiodic; it never repeats itself, and for $A=4$ this aperiodicity becomes quite complicated – in fact, it becomes chaotic.

One important feature of this chaotic behaviour is that it resembles randomness. First, if we simply look at this graph, we do not see any obvious repetitive structure – no periodicity. Secondly, even if we apply some slightly more sophisticated statistical methods to this function, we are not assured of finding any structure. Many chaotic systems pass as random under common statistical tests.

So, what have we learnt? We see that a simple equation such as the one above, which includes a self-regulating mechanism, can produce series that look quite random. But this equation is anything but random. If we know $p(t)$ exactly, then, in principle, we can predict $p(t+1)$ exactly. On the other hand, there is no predictability for a random process, by definition. Therefore, we can have two diametrically opposite underlying mechanisms, one random, and one deterministic, which produce price sequences that look more or less the same.

Now, change gears for a moment and describe, generally, the way in which a market is supposed to work, according to some simple version of received economic wisdom. Taking again the gold market, suppose that the market for gold is highly liquid; many people are buying and selling gold all the time, and the quantity of gold that changes hands is large. Furthermore, suppose that the larger economic environment surrounding the gold market is more or less stable: the international situation is stable, there are no revolutions or upheavals expected, production, energy supplies and international trade are all stable and normal, nothing untoward is happening or is expected to happen. Now suppose that in this idyllic world, one jewellery manufacturer decides to increase output. Naturally, the demand for gold goes up. The market is very liquid, so you would expect it to respond quickly to this increased demand, and prices should rise.

The price of gold should move from one fairly stable value to a higher stable value in the short interval that it takes for the market to adjust to this new demand. Of course, during the adjustment period, the price could wobble up and down as market participants try to take advantage of the change in demand to reap a profit. For example, a rumour might start that the jeweller is about to increase production even more than he or she actually intends. That would send the price up momentarily, but eventually, it would settle down to a new 'correct' and stable value.

This story seems reasonable, and has behind it the basic ideas of the efficient market hypothesis. Very generally, the efficient market hypothesis says that in a public, liquid market, the prices respond quickly to unpredictable changing circumstances and there is, in some sense, no possibility of consistent long-term profits. In particular, changes in the price of some commodity in a liquid market are generated by new information, which the market reacts to quickly. In its simplest version, the new price of gold just reflects the new demand for gold, and because nothing else in the environment is changing, this new price should be stable, and should not change unless new information enters the market. So in a stable economic environment, the price of a commodity should be the same, except possibly for small variations due to new pieces of information entering the market place. Because new information, by its definition, cannot be anticipated, these price movements are also not predictable, that is, they are random. If you look at real price movements in some financial markets, you will see changes in the price that do appear to be random.

This is a very simple view of the efficient market hypothesis. There are more sophisticated versions, but they more or less share the philosophy expressed here. This view has been around for some 25 years, and has a good deal of credibility, at least in academic circles – although defectors from the efficient market camp have recently become more numerous.

But contrast this model with the behaviour we would expect in a market with nonlinear self-regulatory mechanisms. Suppose that the jeweller's increased demand induced a rise in the price. The price could not rise indefinitely; at some point, gold would be too expensive and people would invest their money in other commodities. The self-regulating mechanisms of the market would act to limit the increase in price, and would do so in a nonlinear way.

As we saw in our example of the logistic map, depending on the value of A, the price might, indeed, settle down to a new stable value, but it might also do something different: the nonlinear regulatory mechanisms could create all kinds of interesting price movements as a function of time, even random-looking ones. It may be that in a nonlinear market, the price movements may not be solely due to the new information affecting the market but may result partly from the nonlinear dynamics of the market itself. In fact, we ought to expect that some of the underlying structure of the dynamics of the market will be reflected in the price movements. Of course, there are also many sources of noise and new unpredictable information in the markets. In addition, whatever self-regulating mechanisms exist are vastly more complicated than that of the simple logistic map.

To complicate matters still more, the environment in which a market exists is not static. Changes take place in societies and economies on all time scales, from seconds to millennia. A financial market, coupled to other markets and to the society at large, will, in some enormously complex way, reflect in its prices all of these changes (as well as the anticipation of changes) over all time scales, incorporating new information, and expressing the effects of its own underlying (generally nonlinear) self-regulating mechanisms.

There is a wonderful little example of the way in which feedback and self-regulatory mechanisms can operate, and can produce unexpected results. It is called the beer distribution game. The Sloane School of Management at the Massachusetts Institute of Technology has used it for many years to introduce business students to concepts of economic dynamics. The game is a role-playing game in which players take the role of either a retailer, wholesaler, distributor, or brewery. The game sets up rules whereby the retailers can send orders to the wholesaler, the wholesaler to the distributor, and the distributor to the brewery for a number of cases of beer for each time-step – called a week. The retailer is told each week how many cases of beer consumers are buying that week. Players try to have on hand enough cases to satisfy the demands of their customers, without overstocking. They are penalized for overstocking and understocking the beer. The retailer will try to order just enough beer from the wholesaler to cover the anticipated demand for the next week, the wholesaler will try to do the same by ordering from the distributor, and the distributor will try to order enough beer from the brewery to cover the anticipated demand from the wholesaler. Similarly, the brewery will try to produce just enough beer to meet the demands of the distributor.

The game begins with a constant demand on the retailer by the customers of, say, four cases of beer per week. After a few weeks, retail demand jumps to eight cases per week, and stays there for the rest of the game. You would suppose that the players all the way down the line would smoothly adjust to this increased demand by quickly doubling their orders (or, in the case of the brewery, production) of beer. But that does not happen. Instead, the weekly orders and inventories of the players typically undergo huge oscillations. For example, by about the 30th week, it is quite common to see distributors ordering 40 cases of beer per week from the brewery, even though consumers are buying a total of only eight cases of beer per week. Clearly, the feedback and regulatory dynamics of this extremely simple system produce highly unexpected and erratic behaviour.

Economists have studied other versions of this experimental game with different rules intended to mimic the dynamics of various aspects of the economy. Generally, such games produce outcomes that show the typical features expected in a nonlinear system – limit cycles, multiple periodic phenomena and chaos.

These games are, of course, much simpler than any real economy. It is true that they contain some of the kinds of feedback elements we expect in a real economy. But because real economies are so much more complex, it is possible that the chaotic effects we see in the games may simply not be present in a real economy. The real system may have so many different things going on that all the interesting deterministic effects are just averaged away. We may find ourselves back in a situation in which the economy can be described by linear processes with a lot of noise thrown in. To understand whether this happens, we have to go directly to the real economic data. Unfortunately, real economic data are very difficult to work with. The data sets are fairly short, and there is in any case a lot of noise. Furthermore, many of the statistical methods that economists usually apply will fail to pick up the kinds of nonlinear effects we are looking for. In the past few years, however, new techniques of analysis have been invented that are much more sensitive to the presence of underlying nonlinearities, and can distinguish between certain kinds of randomness and chaos. These new methods, developed by William Brock at the University of Wisconsin and his collaborators, and by some other research groups, including our group at the University of Michigan, are based on the idea that chaotic systems often reveal their structure most clearly when viewed in higher dimensions.

Take, for example, the logistic map discussed above. Looking at the price of gold $p(t)$ in the chaotic regime (see Figure 14.1e), it is certainly not clear whether the values $p(t)$ are random or whether there is some underlying structure and predictability. However, suppose we construct a two-dimensional graph of $p(t)$ in which we plot $p(t+1)$ versus $p(t)$. That is, we take two consecutive numbers from the list $p(t)$. These two numbers are the x and y coordinates of a point in a plane. We plot that point. Then we shift down the list $p(t)$ by one step and take the next pair, and consider those two values to be the x and y coordinates of another point in the plane. We continue this way until we go through the whole series $p(t)$.

Now, if the values $p(t)$ were really random, then our two-dimensional plot would look just like a scatter of points, as shown in Figure 14.2a. On the other hand, in Figure 14.2b, we show the result of the plot constructed from the logistic map. We see very clearly a parabola, which is strikingly different from the scatter of points in Figure 14.2a. The existence of an object with some structure in some dimension, as in Figure 14.2b, is an indication that underlying the random-looking series is a deterministic process. Of course, this example is particularly simple. In more realistic cases, one may have to view the series in more than two dimensions to see any structure at all, and even then, if (as is usually the case) there is noise in the system,

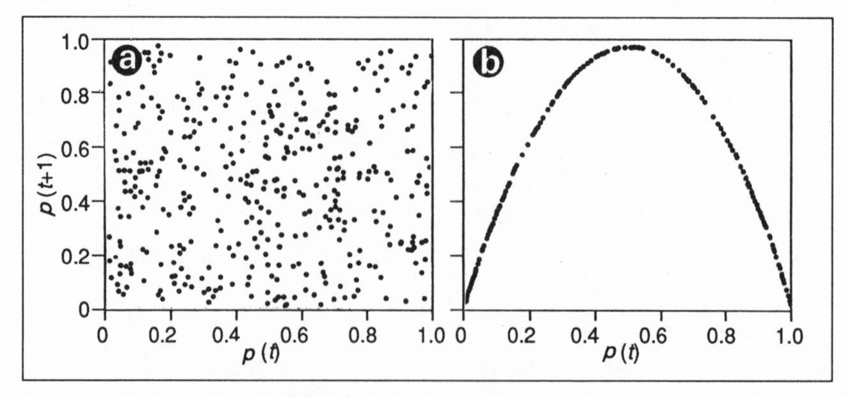

Figure 14.2 Randomness and chaos in gold prices. Compare the random scatter in **a** with the chaotic parabolic plot in **b**.

the structure may be fuzzed out, and difficult to distinguish from the scatter of points that characterizes randomness. It is at this point that the new techniques of statistical analysis are very useful. By properly sampling the higher-dimensional space in which we view the series $p(t)$ it is possible to identify the existence of an underlying deterministic process even in the presence of a good deal of noise.

A fair number of financial and economic data series have been analysed using these methods. Although not all time series show these effects, taken all together, there is significant evidence for the existence of underlying nonlinear processes in economics and finance. Furthermore, we have recently devised methods for learning more about the details of the underlying dynamics, which hold promise for the problem of short-term forecasting and prediction in certain kinds of chaotic systems. Using these methods, we have studied a variety of mathematical examples of chaotic systems, as well as some real financial data. We have been able to ascertain systematically some of the deterministic structure of the underlying process and, in some examples, make relatively accurate predictions of the next term in a time series, even if there is random 'noise'. Of course, as described in earlier chapters, this is a subject of immense interest in many fields, including the physical and biological sciences, as well as the financial markets.

Finally, to return to the scene on the trading floor of an exchange: all those traders acting individually, jockeying for position and profit, are both the observers of the market and the phenomenon of the market, both the audience and the actors. They anticipate changes, assimilate information, make their best guess about the direction of the market, using different techniques and with different motivations (but all wanting to make a profit), and place their bets. They act, and they watch, and they act again. If things get too far out of line they usually (but not always) act in concert to bring prices back into line. Remarkably, the system, for the most part, regulates itself – with a little judicious regulatory help from the government. These markets are enormously complex. Their self-regulatory mechanisms are exquisitely intricate, reflecting the effects of human psychology, social behaviour and, to some extent, rational thought. It is hopeless to try to model such a system in detail. But the prospect of gaining deeper insight into the behaviour of such systems by recognizing their intrinsic nonlinear structure is enormously exciting and promising.

Further reading

ROBERT SAVIT, 'When random is not random: an introduction to chaos in market prices', *J. Future Markets*, Vol. 8, 1988, p. 271.

P. ANDERSON, K. ARROW and D. PINES (eds.) *The Economy as an Evolving Complex System*, Santa Fe Institute Studies in the Sciences of Complexity, Vol. 5, Addison-Wesley, 1988.

15
Quantum physics on the edge of chaos

MICHAEL BERRY

Quantum physics describes the world of the very small. Classical Newtonian physics describes larger scales. But in the border country between the two, rigorous mathematical descriptions are difficult to find, and chaos rears its head.

Before 1900, the foundation of physics was Newtonian mechanics. The main principle was that forces deflect objects according to laws that we can describe using simple mathematics. It was natural to think that because the laws are simple the motion of objects must be simple too. The spectacular success of classical mechanics in explaining regularities in the motion of the Moon and planets encouraged this view. It also inspired the invention of mechanisms imitating those regularities, such as clockwork.

This view is mistaken. Simple deterministic laws can generate very complicated and even random motion, because some systems are so unstable that the course of their trajectories depends sensitively on how they are started off. Even the motion of a billiard ball – the archetypal Newtonian system – can become complicated in certain simple systems. We can idealize the billiard ball as a point reflected elastically, in other words, without any loss of energy, from the boundary of the region in which it moves. Figure 15.1 shows what happens to the ball when we confine it in regions with different shapes. If the enclosure is a rectangle or a circle, the ball bounces round in a regular pattern. But if the boundary is shaped like a stadium or a bulgy 'Africa',

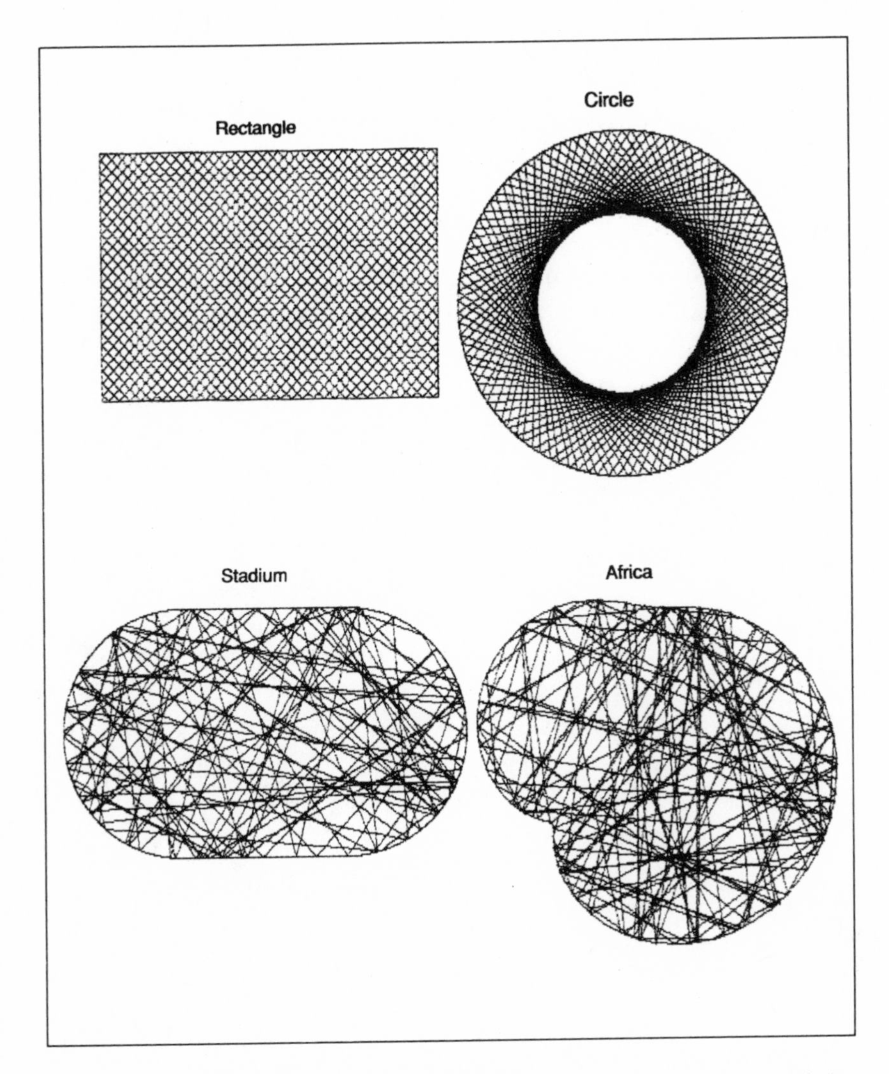

Figure 15.1 The trajectory of a billiard ball depends on the shape of the boundary that confines it. In the rectangle and circle, the orbits are regular; in the stadium and 'Africa', they are chaotic.

then the ball bounces around chaotically, following no regular pattern at all.

Nowadays, the 'chaology' of classical mechanics is an intensively active area of research – chaology is a revival of a term used by theologians two centuries ago to mean the study of what existed before the Creation. It has applications ranging from the irregular tumbling of Saturn's satellite Hyperion to the intricate orbits of food particles in a blender.

Since the 1920s, we have known that Newtonian mechanics, chaotic or not, is but an approximation to deeper truths about physics described by quantum mechanics. When dealing with objects and processes on atomic and smaller scales, it is quantum, not classical, theory that agrees with experiment. In its most familiar form, quantum mechanics is a wave theory. One consequence of this is that the energy of an isolated atomic system cannot take any value, as in classical physics. It is restricted to a set of possible energy levels. A common analogy is with a guitar string, on which waves have a discrete set of frequencies, or harmonics, that depend on the string's length, tension and density. The lowest energy corresponds to the ground state, in which the system usually exists. Higher energies correspond to excited states. Shining light of the appropriate frequency onto an atom will drive it into an excited state.

Quantum physics has its own randomness, to be sharply distinguished from any irregularity that Newtonian trajectories might possess. We cannot, for example, predict when a radioactive nucleus will decay, or where the next photon in a laser beam will strike a screen. But from the equations of quantum mechanics we can calculate with great accuracy the probabilities of these events from the intensities of the quantum waves. So quantum randomness lies not in the waves but in the processes the waves describe.

In the everyday world we can see, the direct effects of quantum mechanics are unobservably small because the wavelengths of the quantum waves are so small. Even for a bacterium, only a thousandth of a millimetre across, creeping at one millimetre an hour, the wavelength is a million times smaller than the bacterium itself – and 100 times smaller than an atom. On scales larger than an atom, we know that Newtonian mechanics works well, so that quantum mechanics must give the same predictions, in spite of its very different conceptual basis.

Niels Bohr, one of the pioneers of quantum physics, saw this relationship between Newtonian and quantum mechanics as a deep truth, which he called 'the correspondence principle'. Quantum mechanics must agree with Newtonian mechanics when applied to large or heavy systems – that is in the 'classical limit' where we can neglect wave effects. We are familiar with the principle applied to optics. Light is a wave, but, in explaining how cameras and telescopes work, it is useful to think in terms of well-defined rays, very similar to the trajectories of the particles, or 'corpuscles', Newton envisaged as the constituents of light.

We know that Newtonian physics can give rise to chaotic behaviour. According to the correspondence principle, quantum physics is identical to Newtonian physics in the classical limit. So how does the quantum system reflect this fact? What features of the way it evolves, and the way its energy levels are distributed, betray the irregularity of the Newtonian trajectories? Can quantum systems become chaotic as they approach the classical limit? These are questions of quantum chaology, an emerging science that is leading to the discovery of unfamiliar regimes of behaviour in microscopic systems.

The first surprise came 10 years ago, in a theoretical study by the Italian-Soviet-American collaboration of Giulio Casati at Milan, Boris Chirikov and Felix Izraelev at Novosibirsk, and Joseph Ford at Atlanta. They investigated how electrons in highly excited atoms – atoms with electrons in states of extremely high energy – absorb energy from radiation shining on them. To avoid tedious computations, necessary for real atoms containing many electrons, they thought of a simple idealized model of a circulating electron as a bead on a circular wire, endlessly pursuing its orbit (see Figure 15.2). The waves of radiation shining on the electron produce an oscillatory force. The researchers represented this as a sequence of impulses that provided a series of kicks to the 'bead'. The strength of the impulses depends on the position of the bead on the circle.

The orbits of this 'kicked rotator', when considered as a Newtonian system, can be regular or chaotic, depending on the strength of the impulses: stronger kicks give more chaos.

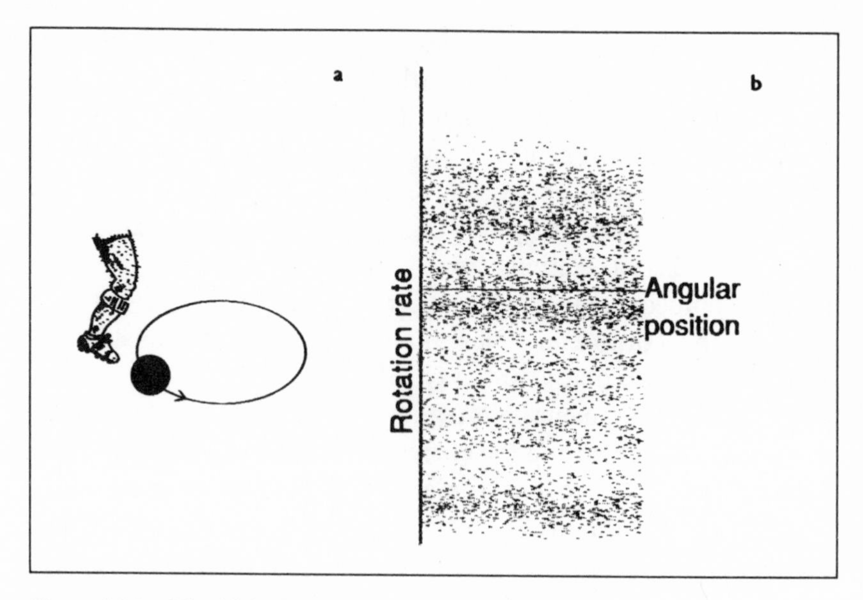

Figure 15.2a The kicked rotator, an idealized model of an excited electron. *Figure* 15.2b The chaotic 'dust' generated by plotting angular position versus the rate of rotation after 5000 kicks.

Classical regularity is a steady state in which the rotator, on the average, gives back as much energy as it absorbs. We also find this behaviour when we carry out the quantum version of the 'experiment', by making the bead so light that we have to take into account its wavelength. Classical chaos, on the other hand, corresponds to erratic diffusion, with the rotator continuing to absorb energy at a rate which, on average, is constant.

In the chaotic case, however, the corresponding quantum rotator behaves differently (see Figure 15.3). For a while, the growth of the rotator's energy follows the classical straight line, but eventually, at a certain 'break-time', the energy begins to grow much more slowly, and may even decrease. This was a surprise. Quantum mechanics has suppressed the classical chaos. At first sight, it looks as though we have a conflict with the correspondence principle. But when we adjust the quantum model to make it more classical, for example by making the particle heavier, the break-time, which signals the onset of non-classical effects, gets later and later. The theorists discovered this suppression of chaos by using a computer to solve the quantum equations numerically. After a decade of study, it is becoming clear that the suppression is a delicate and subtle wave-interference effect, but physicists have not yet worked this out in full detail.

A real atom differs from a model rotator in that an electron can become so excited as to leave the nucleus altogether – thus ionizing the atom. The probability that a given period of exposure to radiation of a certain frequency will result in ionization depends on the intensity of the radiation and how excited the atom is to start with. These conditions determine whether the classical electron trajectories are chaotic or regular, and whether the probabilities of ionization calculated using the classical approximation are the same as those calculated by the more accurate laws of quantum mechanics.

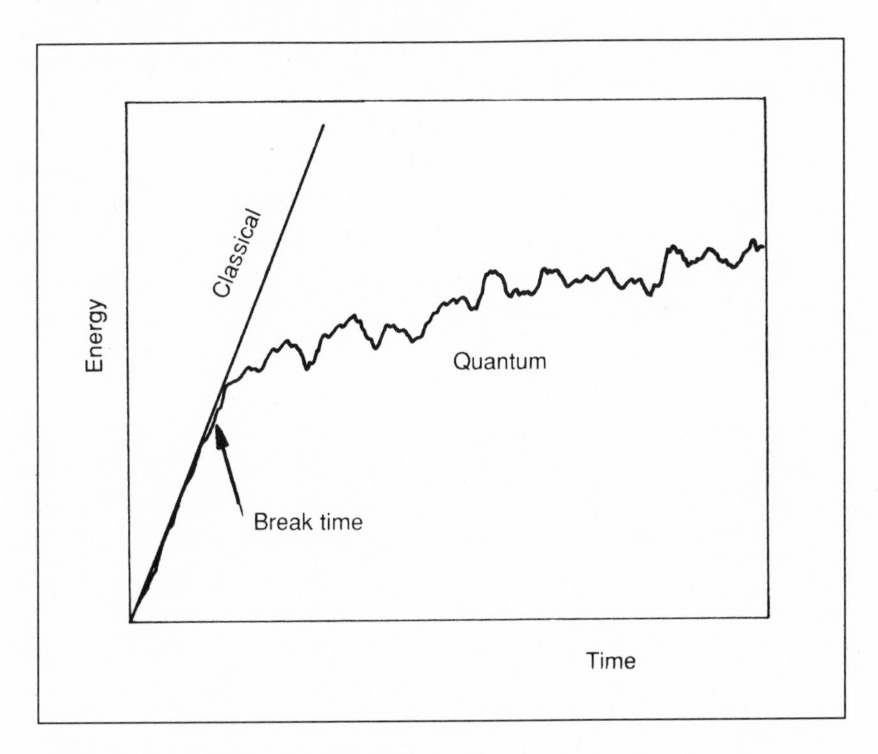

Figure 15.3 After the 'break-time', the quantum rotator absorbs energy more slowly than the chaotic classical rotator.

Physicists have carried out experiments on hydrogen atoms illuminated with microwaves in classically chaotic regimes for which classical and quantum predictions agree – that is before the break-time. They were surprised to find systems as small as atoms behaving classically. Remember the absorption and emission of photons by atoms is a highly non-classical process. The pattern of emission and absorption lines in a spectrum was one of the observations that led to the discovery of quantum mechanics.

The difference between the new spectroscopy and the old is that the new experiments employ intense radiation, and the atoms are in highly excited states to start with, so they absorb and emit large numbers of photons, rather than one or two. Theory predicts that, as with the rotator, quantum mechanics will eventually suppress classical chaos. Experiments to test this important effect would require that we measure the ionization after much longer periods of illumination. This is technically difficult and, so far, no one has managed to do it.

Now we turn to the quantum chaology of systems that are either isolated or else are influenced by external forces that do not vary – in contrast to the oscillatory force of radiation just considered. The energy levels of such systems describe their quantum states. It turns out that the distribution of highly excited quantized energy levels – the pattern of notes of the harmonics on the musical analogy – depends in a fundamental way on whether the trajectories of the corresponding classical system are regular or chaotic.

A system encompassing both extremes is the single electron of a hydrogen atom in a very strong magnetic field – for example, the magnetic field in a white dwarf star, which can be a billion times greater than the Earth's field. At low energies, the nucleus of the hydrogen atom, the proton, binds its electron tightly. The electrostatic force between the proton and the electron completely dominates the magnetic force. The classical orbits are ellipses (see Figure 15.4) like the paths of planets round the Sun, and there is no chaos. At very high energies, the electron is far from the nucleus and now it is the magnetic force that

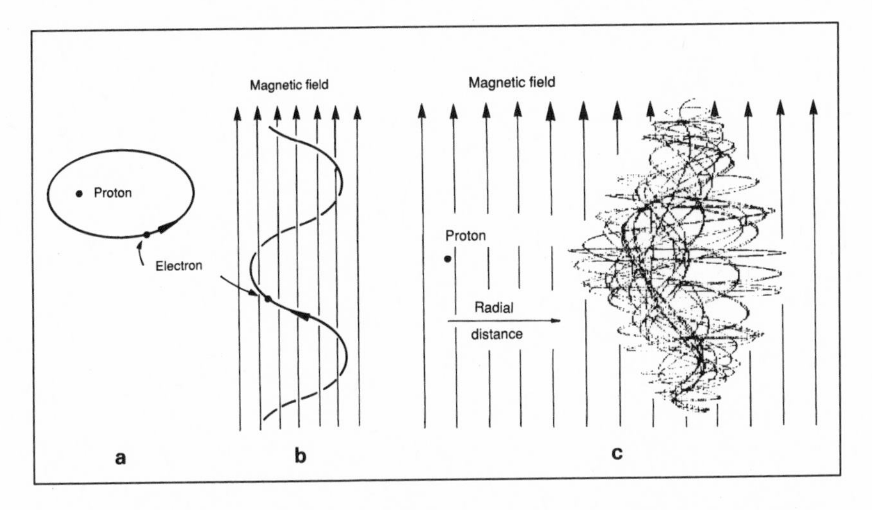

Figure 15.4 Electron orbits: **a**, regular motion in the field of a proton alone; **b**, regular motion in a magnetic field alone; **c**, chaotic motion in a combined field.

dominates. The orbits are helices spiralling round the lines of the magnetic field, and again there is no chaos. At intermediate energies, however, the two forces are comparable but exert contrary influences. The classical electron resolves the contradiction by moving chaotically.

For the electrons behaving in a quantum fashion, we have to compare the distributions of large numbers of excited energy levels in the regular and chaotic regimes. One way to do this is by computing the statistics of the levels. One convenient statistic is the spacings between neighbouring levels, calculated at low and intermediate energies in the atom's spectrum of energy levels. If the levels are regularly arranged, like the rungs of a ladder, the distribution of spacings will cluster about the average spacing, producing a curve like that in Figure 15.5a. You would get a similar distribution by plotting the heights of a group of people. In this case there are few small spacings – it is as though the levels repel each other. If, on the other hand, the levels are randomly distributed – that is, uncorrelated, like the arrival times of raindrops in a shower – the distribution of spacings will be broad, with a preponderance of small spacings. The surprise

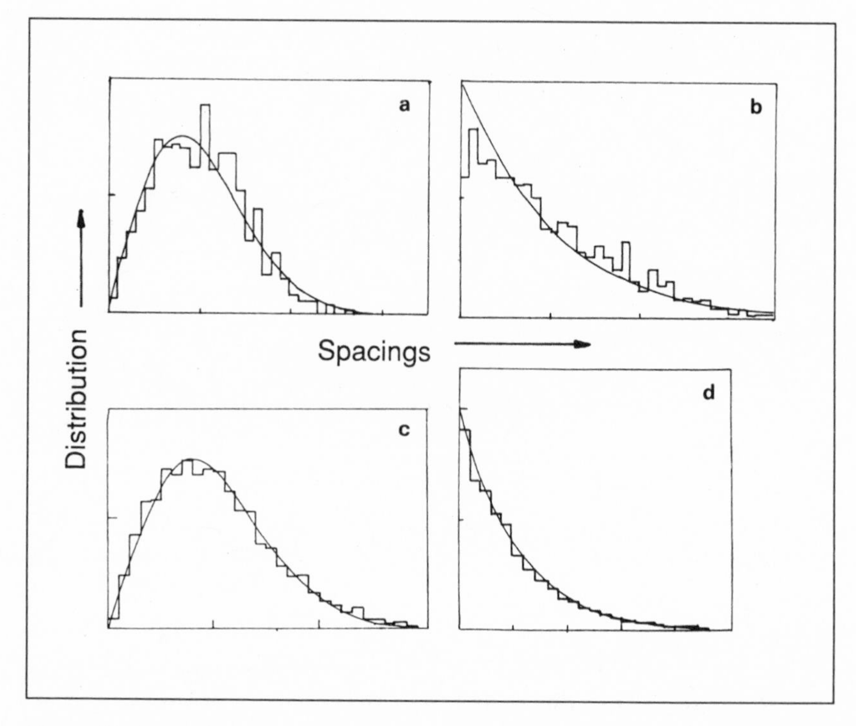

Figure 15.5 The distribution of spacings between neighbouring energy levels: **a**, hydrogen in a magnetic field with chaotic classical orbits showing repulsion of levels; **b**, hydrogen in a magnetic field with regular classical orbits showing clustering of levels; **c**, the energy levels for quantum billiards in a stadium; **d**, the energy levels for quantum billiards in a rectangle.

this time was the discovery that the levels are more regularly arranged when the classical orbits are chaotic (see Figure 15.5a), and randomly distributed when the orbits are regular (see Figure 15.5b). Experiments on magnetized hydrogen confirm even the fine details of the theoretically calculated spectrum.

This behaviour is not just a peculiarity of the magnetized hydrogen atom. On the contrary, the spacings of the quantum energy levels always depend only on whether the classical orbits are chaotic or regular and not on any other details of the system. To illustrate this, Figure 15.5 also shows the distributions of the spacings of energy levels for the quantized versions of two of the billiard ball games in Figure 15.1. The stadium game is classically chaotic but has regular spacings of its quantum levels (see Figure

15.5c), while the rectangle game is classically regular but has random spacings of its quantum levels corresponding to chaotic motion. These tend to repel one another. Quantum billiards might appear to be an exotic creation of theorists, far removed from reality, but exactly the same mathematics describes the frequencies of a vibrating membrane shaped like the billiard table. In the three-dimensional version, it also describes the acoustics of a concert hall.

The repulsion of levels in Figures 15.5b and 15.5d is not the most general quantum signature of classical chaos, because all systems so far discussed have a special feature, namely symmetry. The atom in a magnetic field has the symmetry of a cylinder, and the movement of the billiards is symmetrical with respect to time, in the sense that, if at any instant the velocity of the moving billiard ball reverses, it will retrace its previous path. When there is no symmetry of any kind and the classical orbits are chaotic – a combination of circumstances still out of range of experiment – theory predicts that repulsion between the levels remains but it is of the slightly stronger kind shown in Figure 15.6a, in which

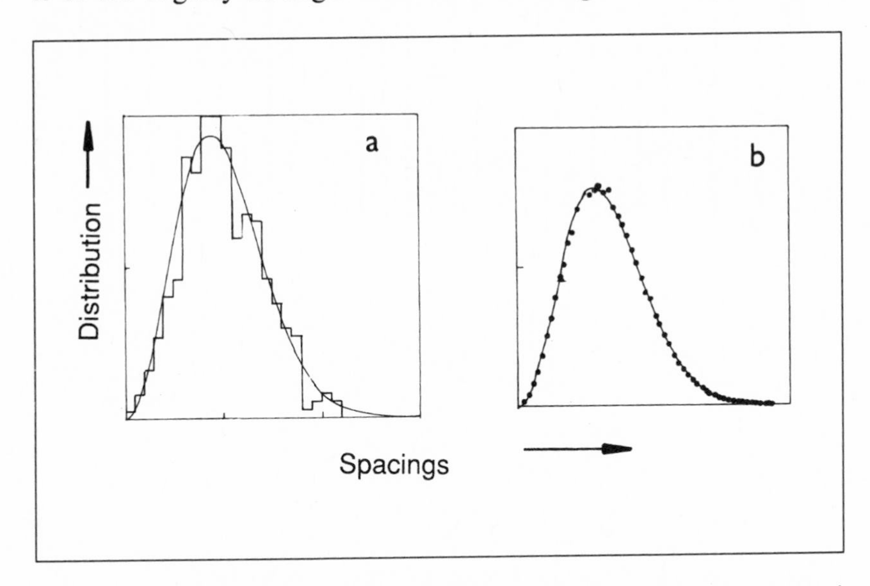

Figure 15.6 The distribution of spacings where there is no symmetry: **a**, several hundred energy levels of the 'Africa' game of quantum billiards in a magnetic field; **b**, 100 000 zeros of Riemann's zeta function.

the slope of the curve vanishes at zero spacing – the curve flattens out. This particular calculation was for the energies of a charged quantum particle moving in the 'Africa' billiard table of Figure 15.1 with a magnetic field acting at right angles to the plane, but it represents quantum chaology in the most general case.

At this point, quantum chaology makes unexpected contact with one of the long-standing problems of pure mathematics, namely the Riemann hypothesis of number theory. In 1859, Georg Bernhard Riemann – a German mathematician who also developed the study of geometry to include that with more than three dimensions – was studying the distribution of prime numbers. He devised a quantity, which he called the zeta function, whose value depends on position in a plane of complex numbers. Complex numbers (denoted by s) have a 'real' part and an 'imaginary' part involving the square root of minus 1. The x axis represents real numbers and the y axis imaginary numbers. Riemann's function was the extension to the whole s plane of

$$\text{zeta}\,(s) = 1 + \tfrac{1}{2}^{s} + \tfrac{1}{3}^{s} + \ldots$$

His famous hypothesis was that the points at which the zeta function vanishes – its zeros – lie on the straight line with $x = \tfrac{1}{2}$. (If the hypothesis were true, certain theorems about prime numbers would follow.) Numerical studies have shown that the first 1 500 000 000 zeros lie on Riemann's line, but nobody has been able to prove that they all do (see Figure 15.7).

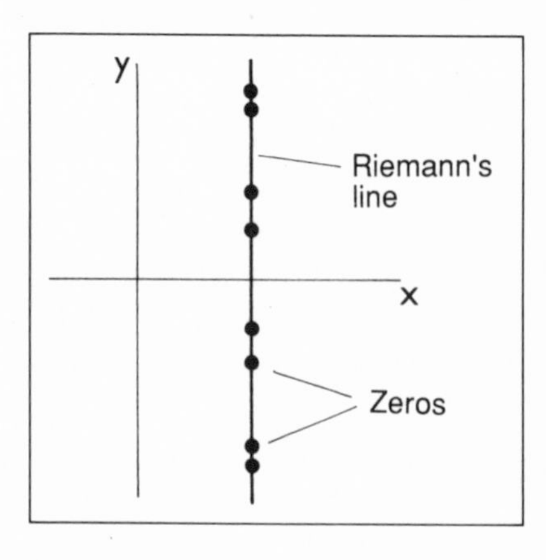

Figure 15.7 The plane inhabited by Riemann's zeta function. His hypothesis is that the function's zeros lie on the line shown.

The connection with quantum chaology comes in calculations by Andrew Odlyzko (A T & T Bell Laboratories, New Jersey) of the distribution of the spacings between neighbouring zeros; this takes 20 hours of Cray supercomputer time and results in the graph in Figure 15.6. It is not just the evident similarity between Figures 15.6a and 15.6b, but also a variety of other evidence, which suggests that underlying Riemann's zeta function is some unknown classical, mechanical system whose trajectories are chaotic and without symmetry, with the property that, when quantized, its allowed energies are the Riemann zeros. These connections between the seemingly disparate worlds of quantum mechanics and number theory are tantalizing.

The phenomena of quantum chaology lie in the largely unexplored border country between quantum and classical mechanics; they are part of semiclassical mechanics. This is an area where rigorous mathematical development, as employed elsewhere in mechanics, is difficult. Most discoveries have been made by computer experiments with the quantum equations, guided by intuition and analogy. As the subject matures we can expect, on the one hand, more experiments on real physical systems, and, on the other, the precise formulation and proof of mathematical theorems.

Further reading

MARTIN C. GUTZWILLER, 'Quantum Chaos', *Scientific American,* January 1992, p. 78.

RODERICK V. JENSON, 'Quantum Chaos', *Nature,* 23 January 1992, p. 311.

16

A random walk in arithmetic

GREGORY CHAITIN

God plays dice not only in physics but also in pure mathematics. Mathematical truth is sometimes nothing more than a perfect coin toss.

The notion of randomness obsesses physicists today. To what extent can we predict the future? Does it depend on our own limitations? Or is it in principle impossible to predict the future? The question of predictability has a long history in physics. In the early 19th century, the classical deterministic laws of Isaac Newton led Pierre Simon de Laplace to believe that the future of the Universe could be determined forever.

Then quantum mechanics came along. This is the theory that is fundamental to our understanding of the nature of matter. It describes very small objects, such as electrons and other fundamental particles. One of the controversial features of quantum mechanics was that it introduced probability and randomness at a fundamental level to physics. This greatly upset the great physicist Albert Einstein, who said that God did not play dice.

Then, surprisingly, the modern study of nonlinear dynamics showed us that even the classical physics of Newton had randomness and unpredictability at its core. The theory of chaos has revealed how the notion of randomness and unpredictability is beginning to look like a unifying principle.

It seems that the same principle even extends to mathematics. I can show that there are theorems connected with number theory that cannot be proved because when we ask the appropriate questions, we obtain results that are equivalent to the random toss of a coin.

My results would have shocked many 19th-century mathematicians, who believed that mathematical truths could always be proved. For example, in 1900, the mathematician David Hilbert gave a famous lecture in which he proposed a list of 23 problems as a challenge to the new century. His sixth problem had to do with establishing the fundamental universal truths, or axioms, of physics. One of the points in this question concerned probability theory. To Hilbert, probability was simply a practical tool that came from physics; it helped to describe the real world when there was only a limited amount of information available.

Another question he discussed was his tenth problem, which was connected with solving so-called 'diophantine' equations, named after the Greek mathematician Diophantus. These are algebraic equations involving only whole numbers, or integers. Hilbert asked: 'Is there a way of deciding whether or not an algebraic equation has a solution in whole numbers?'

Little did Hilbert imagine that these two questions are subtly related. This was because Hilbert had assumed something that was so basic to his thinking that he did not even formulate it as a question in his talk. That was the idea that every mathematical problem has a solution. We may not be bright enough or we may not have worked long enough on the problem but, in principle, it should be possible to solve it – or so Hilbert thought. For him, it was a black or white situation.

It seems now that Hilbert was on shaky ground. In fact, there is a connection between Hilbert's sixth question dealing with probability theory and his tenth problem of solving algebraic equations in whole numbers that leads to a surprising and rather chilling result. That is: randomness lurks at the heart of that most traditional branch of pure mathematics, number theory.

Clear, simple mathematical questions do not always have clear answers. In elementary number theory, questions involving diophantine equations can give answers that are completely random and look grey, rather than black or white. The answer is random because the only way to prove it is to postulate each answer as an additional independent axiom. Einstein would be horrified to discover that God plays dice not only in quantum and classical physics but also in pure mathematics.

Where does this surprising conclusion come from? We have to go back to Hilbert. He said that there should be a mechanical procedure to decide whether a mathematical proof is correct or not, and the axioms should be consistent and complete. If the system of axioms is consistent, it means that you cannot prove both a result and its contrary. If the system is complete, then you can also prove any assertion to be true or false. It follows that a mechanical procedure would ensure that all mathematical assertions can be decided mechanically.

There is a colourful way to explain how this mechanical procedure works: the so-called 'British Museum algorithm'. What you do – it cannot be done in practice because it would take forever – is to use the axiom system, set in the formal language of mathematics, to run through all possible proofs, in order of their size and lexicographic order. You check which proofs are correct – which ones follow the rules and are accepted as valid. In principle, if the set of axioms is consistent and complete, you can decide whether any theorem is true or false. Such a procedure means that a mathematician no longer needs ingenuity or inspiration to prove theorems. Mathematics becomes mechanical.

Of course, mathematics is not like that. Kurt Gödel, the Austrian logician, and Alan Turing, the father of the computer, showed that it is impossible to obtain both a consistent and complete axiomatic theory of mathematics and a mechanical procedure for deciding whether an arbitrary mathematical assertion is true or false, or is provable or not.

Gödel was the first to devise the ingenious proof, couched in number theory, of what is called the incompleteness theorem. But I think that the Turing version of the theorem is more fundamental and easier to understand. Turing used the language of the computer – the instructions, or program, that a computer needs to work out problems. He showed that there is no mechanical procedure for deciding whether an arbitrary program will ever finish its computation and halt.

To show that the so-called halting problem can never be solved, we set the program running on a Turing machine, which is a mathematical idealization of a digital computer with no time limit. (The program must be self-contained with all its data wrapped up inside the program.) Then we simply ask: 'Will the program go on forever, or at some point will it say "I'm finished" and halt?'

Turing showed that there is no set of instructions that you can give the computer, no algorithm, that will decide if a program will ever halt. Gödel's incompleteness theorem follows immediately, because if there is no mechanical procedure for deciding the halting problem, then there is no complete set of underlying axioms either. If there were, they would provide a mechanical procedure for running through all possible proofs to show whether programs halt – although it would take a long time, of course.

To obtain my result about randomness in mathematics, I simply take Turing's result and just change the wording. What I get is a sort of a mathematical pun. Although the halting problem is unsolvable, we can look at the probability of whether a randomly chosen program will halt. We start with a thought experiment using a general purpose computer that, given enough time, can do the work of any computer – the universal Turing machine.

Instead of asking whether or not a specific program halts, we look at the ensemble of all possible computer programs. We assign to each computer program a probability that it will be chosen. Each bit of information in the random program is chosen by tossing a coin, an independent toss for each bit, so that a program containing so many bits of information, say N bits, will have a probability of 2^{-N}. We can now ask what is the total probability that those programs will halt. This halting probability, call it Ω, wraps up Turing's question of whether a program halts into one number between 0 and 1. If the program never halts, Ω is 0; if it always halts, Ω is 1.

In the same way that computers express numbers in binary notation, we can describe Ω in terms of a string of 1s and 0s. Can we determine whether the Nth bit in the string is a 0 or a 1? In other words, can we compute Ω? Not at all. In fact, I can show that the sequence of 0s and 1s is random using what is called algorithmic information theory. This theory ascribes a degree of order in a set of information or data according to whether there is an algorithm that will compress the data into a briefer form.

For example, a regular string of 1s and 0s describing some data such as 0101010101 ... which continues for 1000 digits can be encapsulated in a shorter instruction 'repeat 01 500 times'. A completely random string of digits cannot be reduced to a shorter program at all. It is said to be algorithmically incompressible.

My analysis shows that the halting probability is algorithmically random. It cannot be compressed into a shorter program. To get N bits of the number out of a computer, you need to put in a program at least N bits long. Each of the N bits of Ω is an irreducible independent mathematical fact, as random as tossing a coin. For example, there are as many 0s in Ω as 1s. And knowing all the even bits does not help us to know any of the odd bits.

My result that the halting probability is random corresponds to Turing's assertion that the halting problem is undecidable. It has turned out to provide a good way to give an example of randomness in number theory, the bedrock of mathematics. The key was a dramatic development about five years ago. James Jones of the University of Calgary in Canada and Yuri Matijasevic of the Steklov Institute of Mathematics in Leningrad discovered a theorem proved by Edouard Lucas in France a century ago. The theorem provides a particularly natural way to translate a universal Turing machine into a universal diophantine equation that is equivalent to a general purpose computer.

I thought it would be fun to write it down. So with the help of a large computer I wrote down a universal-Turing-machine equation. It had 17 000 variables and went on for 200 pages.

The equation is of a type that is referred to as 'exponential diophantine'. All the variables and constants in it are non-negative integers, 0, 1, 2, 3, 4, 5, and so on. It is called 'exponential' because it contains numbers raised to an integer power. In normal diophantine equations the power has to be a constant. In this equation, the power can be a variable. So in addition to having X^3, it also contains X^Y.

To convert the assertion that the halting probability Ω is random into an assertion about the randomness of solutions in arithmetic, I need only to make a few minor changes in this 200-page universal-Turing-machine diophantine equation. The result, my equation exhibiting randomness, is also 200 pages long. The equation has a single parameter, the variable N. For any particular value of this parameter, I ask the question: 'Does my equation have a finite or infinite number of whole-number solutions?' Answering this question turns out to be equivalent to calculating the halting probability. The answer 'encodes' in arithmetical language whether the Nth bit of Ω is a 0 or a 1. If the Nth bit of Ω is a 0, then my equation for that particular value of N has a finite number of solutions. If the Nth bit of the halting probability Ω is a 1, then this equation for that value of the parameter N has an infinite number of solutions. Just as the Nth bit of Ω is random – an independent, irreducible fact like tossing a coin – so is deciding whether the number of solutions of my equation is finite or infinite. We can never know.

To find out whether the number of solutions is finite or infinite in particular cases, say for k values of the parameter N, we would have to postulate the k answers as k additional independent axioms. We would have to put in k bits of information into our system of axioms, so we would be no further forward. This is another way of saying that the k bits of information are irreducible mathematical facts.

I have found an extreme form of randomness, of irreducibility, in pure mathematics – in a part of elementary number theory associated with the name of Diophantus and which goes back 2000 years to classical Greek mathematics. Hilbert believed that mathematical truth was black or white, that something was either true or false. I think that my work makes things look grey, and that mathematicians are joining the company of their theoretical physics colleagues. I do not think that this is necessarily bad. We have seen that in classical and quantum physics, randomness and unpredictability are fundamental. I believe that these concepts are also found at the very heart of pure mathematics.

Further reading

G. J. CHAITIN, *Information, Randomness and Incompleteness* (2nd edn), World Scientific, Singapore, 1990.

G. J. CHAITIN, *Algorithm Information Theory*, Cambridge University Press, 1990.

17

Chaos, entropy and the arrow of time

PETER COVENEY

The theory of chaos uncovers a new 'uncertainty principle' that governs how the real world behaves. It also explains why time goes in only one direction.

The nature of time is not only central to our understanding of the world around us, including the physics of how the Universe came into being and how it evolves, but it also affects issues such as the relation between science, culture and human perception. Yet scientists still do not have an easily understandable definition of time.

The problem is that, in the everyday world, time appears to go in one direction – it has an arrow. Cups of tea cool, snowmen melt and bulls wreak havoc in china shops. We never see the reverse processes. This unrelenting march of time is captured in thermodynamics, the science of irreversible processes. But underpinning thermodynamics are the supposedly more fundamental laws of the Universe – laws of motion given by Newtonian and quantum mechanics. The equations describing these laws do not distinguish between past and future. Time appears to be a reversible quantity with no arrow.

So we have a conflict between irreversible laws of thermodynamics and the reversible mechanical laws of motion. Does the notion of an arrow of time have to be given up, or do we need to change the fundamental dynamical laws? Today, the theory of chaos can help with the answer.

The second law of thermodynamics is, according to Arthur Eddington, the 'supreme law of Nature'. It arose from a simple observation: in any macroscopic mechanical process, some or all of the energy always gets dissipated as heat. Just think of rubbing your hands together. In 1850, when Rudolf Clausius, the German physicist, first saw the far-reaching ramifications of this mundane observation, he introduced the concept of 'entropy' as a quantity that relentlessly increases because of this heat dissipation. Because heat is the random movement of the individual particles that make up the system, entropy has come to be interpreted as the amount of disorder the system contains. It provides a way of connecting the microscopic world, where Newtonian and quantum mechanics rule, with the macroscopic laws of thermodynamics.

For isolated systems that exchange neither energy nor matter with their surroundings, the entropy continues to grow until it reaches its maximum value at what is called thermodynamic equilibrium. This is the final state of the system when there is no change in the macroscopic properties – density, pressure, and so on – with time. The concept of equilibrium has proved of great value in thermodynamics. Unfortunately, as a result, most scientists talk about thermodynamics and entropy only in terms of equilibrium states, even though this amounts to a major restriction, as we shall soon see.

It is rare to encounter truly isolated systems. They are more likely to be 'closed' (exchanging energy but not matter with the surroundings) or 'open' (exchanging both energy and matter). Imagine compressing a gas in a cylinder with a piston. The gas and the cylinder constitute a closed system, so we have to take into account the entropy changes arising from the exchange of energy with the surroundings as well as the entropy change within the gas.

The traditional thermodynamic approach to describing what happens is as follows. The total entropy of the system and the surroundings will be at a maximum for an equilibrium state. This entropy will not change as the volume of the gas is reduced, provided that the gas and the surroundings remain at all instants in equilibrium. This process would be reversible. In order to achieve it, the difference between the external pressure and that of the gas must be infinitesimally small to maintain the state of equilibrium at every moment. In practice, of course, this so-called 'quasi-static' compression would take an eternity to perform.

The remarkable conclusion is that equilibrium thermodynamics cannot, therefore, describe change, which is the very means by which we are aware of time. The reason why physicists and chemists rely on equilibrium thermodynamics so much is that it is mathematically easy to use: it produces the quantities, such as entropy, describing the final equilibrium state of an evolving system. Entropy is a so-called thermodynamic 'potential'.

In reality, all processes take a finite time to happen and, therefore, always proceed out of equilibrium. Theoretically, a system can only aspire to reaching equilibrium, it will never actually reach it. It is, therefore, somewhat ironic that thermodynamicists have focused their attention on the special case of thermodynamic equilibrium. For the difference between equilibrium and non-equilibrium is as stark as that between a journey and its destination, or the words of this sentence and the full stop that ends it. It is only by virtue of irreversible non-equilibrium processes that a system reaches a state of equilibrium. Life itself is a non-equilibrium process: ageing is irreversible. Equilibrium is reached only at death, when a decayed corpse crumbles into dust.

Obviously, you have to use non-equilibrium thermodynamics when dealing with systems that are prevented from reaching equilibrium by external influences, say where there is a continuous exchange of materials and energy with the environment. Living systems are typical examples.

As an example of a non-equilibrium system, consider an iron rod whose ends are initially at different temperatures. Normally, if one end is hotter than the other, the temperature gradient along the rod would cause the hot end to cool down and the cooler end to warm up until the rod attained a uniform temperature. This is the equilibrium situation. If, however, we maintain one end at a higher temperature, the rod experiences a continual thermodynamic force – a temperature gradient – causing the heat flow, or thermodynamic flux, along the rod. The rod's entropy production is given by the product of the force and the flux, in other words the heat flow multiplied by the temperature gradient.

If the system is close to equilibrium, the fluxes depend in a simple, linear way on the forces: if the force is doubled, then so is the flux. This is linear thermodynamics, which was put on a firm footing by Lars Onsager of Yale University during the 1930s. At equilibrium, the forces vanish and so too do the fluxes.

Ilya Prigogine, a theoretical physicist and physical chemist at the University of Brussels, was among the first to tackle entropy in non-equilibrium thermodynamics. In 1945, he showed that for systems close to equilibrium, the thermodynamic potential is the rate at which entropy is produced by the system; this is called 'dissipation'. Prigogine came up with a theorem of minimum entropy production, which predicts that such systems evolve to a steady state that minimizes the dissipation. This is reminiscent of equilibrium thermodynamics; the final state is uniform in space and does not vary with time.

Prigogine's minimum entropy production theorem is an important result. Together with his colleague Paul Glansdorff and others in Brussels, Prigogine then set out to explore systems maintained even further away from equilibrium, where the linear law for force and flux breaks down, to see whether it was possible to extend his theorem into a general criterion that would work for nonlinear, far-from-equilibrium situations.

Over a period of some 20 years, the research group at Brussels elaborated a theory widely known as 'generalized thermodynamics' (a term, in fact, never used by this group). To apply thermodynamic principles to far-from-equilibrium problems, Glansdorff and Prigogine assumed that such systems behave like a good-natured patchwork of equilibrium systems. In this way, entropy and other thermodynamic quantities depend, as before, on variables such as temperature and pressure.

The Glansdorff-Prigogine criterion makes a general statement about the stability of far-from-equilibrium steady states. It says that they may become unstable as they are driven further from equilibrium: there may arise a crisis, or bifurcation point, at which the system prefers to leave the steady state, evolving instead into some other stable state (see Figure 17.1a).

The important new possibility is that beyond the first crisis point, highly organized states can suddenly appear. In some non-equilibrium chemical reactions, for example, regular colour changes start to happen, so producing 'chemical clocks'; in others, beautiful scrolls of colour arise. Such dynamical states are not associated with minimal entropy production by the system; however, the entropy produced is exported to the external environment.

As a result, we have to reconsider associating the arrow of time with uniform degeneration into randomness – at least on a local level. At the 'end' of time – at equilibrium – randomness may have the last laugh. But over shorter timescales, we can witness the emergence of exquisitely ordered structures which exist as long as the flow of matter and energy is maintained – as illustrated by ourselves, for example.

The Glansdorff-Prigogine criterion is not a universal guiding principle of irreversible evolution, because there is an enormous range of possible behaviours available far from equilibrium.

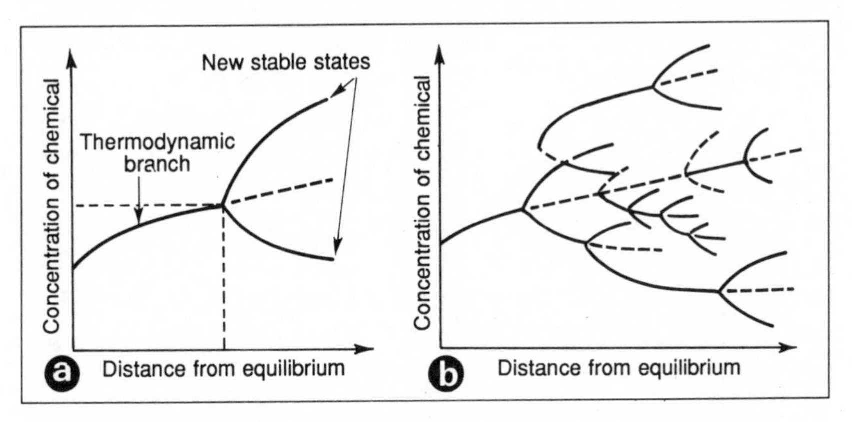

Figure 17.1 Systems far from equilibrium can split into two stable states, as in **a**. As the distance from equilibrium is increased even further, more and more stable states become possible, as in **b** (dashed lines indicate unstable states).

How a non-equilibrium system evolves over time can depend very sensitively on the system's microscopic properties – the motions of its constituent atoms and molecules – and not merely on large-scale parameters such as temperature and pressure. Far from equilibrium, the smallest of fluctuations can lead to radically new behaviour on the macroscopic scale. A myriad of bifurcations can carry the system in a random way into new stable states (see Figure 17.1b). These non-uniform states of structural organization, varying in time or space (or both), were dubbed 'dissipative structures' by Prigogine; the spontaneous development of such structures is known as 'self-organization'.

Existing thermodynamic theory cannot throw light on the behaviour of non-equilibrium systems beyond the first bifurcation point as we leave equilibrium behind. We can explore such states theoretically only by considering the dynamics of the systems. To describe the one-way evolution of such non-equilibrium systems, we must construct mathematical models based on equations that show how various observable properties of a system change with time. In agreement with the second law of thermodynamics, such sets of equations describing irreversible processes always contain the arrow of time.

Chemical reactions provide typical examples of how this works. We can describe how fast such chemical reactions go as they are driven in the direction of thermodynamic equilibrium in terms of rate laws written in the form of differential equations. The quantities that we can measure are the concentrations of the chemicals involved and the rates at which they change with time.

We do not expect to see self-organizing processes in every chemical reaction maintained far from equilibrium. But we often find that the mechanism underlying a reaction leads to differential equations that are nonlinear. Nonlinearities arise, for example, when a certain chemical present enhances (or suppresses) its own production; and they can generate unexpected complexity. Indeed, such nonlinearities are necessary, but not sufficient, for self-organized structures, including deterministic chaos, to appear. An example is the famous Belousov-Zhabotinskii reaction discovered in the 1950s, described by Stephen Scott in Chapter 9.

Today, as other chapters in this book admirably show, many scientists are using nonlinear dynamics to model a dizzying range of complicated phenomena, from fluid dynamics, through chemical and biochemical processes, to genetic variation, heart beats, population dynamics, evolutionary theory and even into economics. The two universal features of all these different phenomena are their irreversibility and their nonlinearity. Deterministic chaos is only one possible consequence; the other is a more regular self-organization; indeed, chaos is just a special, but very interesting, form of self-organization in which there is an overload of order.

We can again ask what is the origin of the irreversibility enshrined within the second law of thermodynamics. The traditional reductionist view is that we should seek the explanation on the basis of the reversible mechanical equations of motion. But, as the physicist Ludwig Boltzmann discovered, it is not possible to base the arrow of time directly on equations that ignore it. His failed attempt to reconcile microscopic mechanics with the second law gave rise to the 'irreversibility paradox' that I mentioned at the beginning of this chapter.

The standard way of attempting to derive the equations employed in non-equilibrium thermodynamics starts from the equations of motion, whether classical or quantum mechanical, of the individual particles making up the system, which might, for example, be a gas. Because we cannot know the exact position and velocity of every particle, we have to turn to probability theory – statistical methods – to relate the average behaviour of each particle to the overall behaviour of the system. This is called statistical mechanics. The approach works very well because of the exceedingly large numbers of particles involved (of the order of 10^{24}).

The reason for using probabilistic methods is not merely the practical difficulty of being unable to measure the initial positions and velocities of the participating particles. Quantum mechanics predicts these restrictions as a consequence of Heisenberg's uncertainty principle. But the same is also true for sufficiently unstable chaotic classical dynamical systems. Ian Percival, in Chapter 1, explained that one of the characteristic features of a chaotic system is its sensitivity to the initial conditions: the behaviour of systems with different initial conditions, no matter how similar, diverges exponentially as time goes on. To predict the future, you would have to measure the initial conditions with literally infinite precision – a task impossible in principle as well as in practice. Again, this means we have to rely on a probabilistic description even at the microscopic level.

Chaotic systems are irreversible in a spectacular way, so we would like to find an entropy-like quantity associated with them because it is entropy that measures change and furnishes the arrow of time. Theorists have made the greatest progress in a class of dynamical systems called ergodic systems. An ergodic system is one which will pass through every possible dynamical state compatible with its energy. The foundations of ergodic theory were laid down by John von Neumann, George Birkhoff, Eberhard Hopf and Paul Halmos during the 1930s and more recently developed by Soviet mathematicians including Andrei Kolmogorov, Dmitrii Anosov, Vladimir Arnold and Yasha Sinai. Their work has revealed a whole hierarchy of behaviours within dynamical systems – some simple, some complex, some paradoxically simple and complex at the same time.

As with the systems described in previous articles on chaos, we can use 'phase portraits' to show how an ergodic system behaves. But in this case, we portray the initial state of the system as a bundle of points in phase space, rather than a single point. Figure 17.2a shows a non-ergodic system: the bundle retains its shape and moves in a periodic fashion over a limited portion of the space. Figure 17.2b shows an ergodic system; the bundle maintains its shape but now roves around all parts of the space. In Figure 17.2c the bundle, whose volume must remain constant, spreads out into ever finer fibres, like a drop of ink spreading in water; eventually it invades every part of the space. This is a consequence of what is called Liouville's theorem. In other words, the total probability must be conserved (and add up to

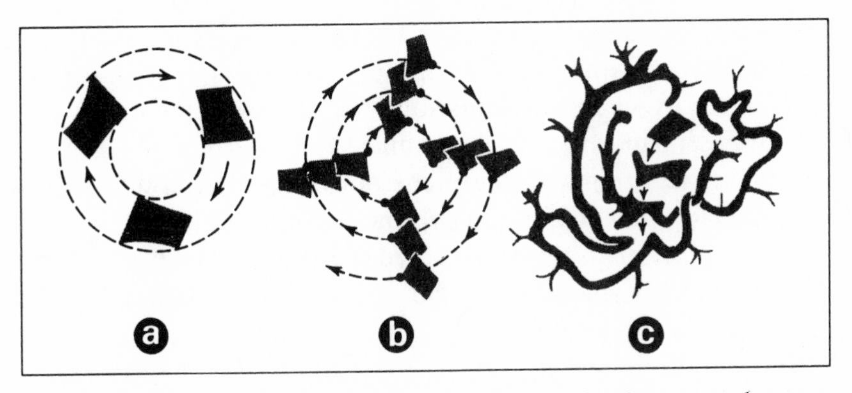

Figure 17.2 The remarkable differences in behaviour in phase space between a simple system **a**, a so-called ergodic system **b**, and a mixing ergodic system **c**, which is chaotic.

1); the bundle behaves like an incompressible fluid drop. This is an example of a 'mixing ergodic flow', and manifests an approach to thermodynamic equilibrium when the time evolution ceases. Such spreading out implies a form of dynamical chaos. The bundle spreads out because all the trajectories that it contains diverge from each other exponentially fast. Hence it can arise only for a chaotic dynamical system.

Mixing flows are only one member of a hierarchy of increasingly unstable and thus chaotic ergodic dynamical systems. Even more random are the so-called *K*-flows, named after Kolmogorov. Their behaviour is close to the limit of total unpredictability: they have the remarkable property that even an infinite number of prior measurements cannot predict the outcome of the very next measurement.

My colleague Baidyanath Misra, working in collaboration with Prigogine, has found an entropy-like quantity with the desired property of increasing with time in this class of highly chaotic systems. The chaotic *K*-flow property is widespread among systems where collisions between particles dominate the dynamics, from those consisting of just three billiard balls in a box (as shown by Sinai in his pioneering work of 1962) to gases containing many particles considered as hard spheres. Many theorists believe, although they have not proved it, that most systems found in everyday life are also *K*-flows. My colleague,

Oliver Penrose of Heriot-Watt University, and I are trying to establish, by mathematically rigorous methods, whether we can formulate exact kinetic equations for such systems in the way originally proposed by Boltzmann.

In joint research with Maurice Courbage, Misra and Prigogine discovered a new definition of time for *K*-flows consistent with irreversibility. This quantity, called the 'internal time', represents the age of a dynamical system. You can think of the age as reflecting a system's irreversible thermodynamic aspects, while the description held in Newton's equations for the same system portrays purely reversible dynamical features.

Thermodynamics and mechanics have been pitted against one another for more than a century, but now we have revealed a fascinating relationship. Just as with the uncertainty principle in quantum mechanics, where knowing the position of a particle accurately prevents us from knowing its momentum and vice versa, we now find a new kind of uncertainty principle that applies to chaotic dynamical systems. This new principle shows that complete certainty of the thermodynamic properties of a system (through knowledge of its irreversible age) renders the reversible dynamical description meaningless, whilst complete certainty in the dynamical description similarly disables the thermodynamic view.

Understanding dynamical chaos has helped to sharpen our understanding of the concept of entropy. Entropy turns out to be a property of unstable dynamical systems, for which the cherished notion of determinism is overturned and replaced by probabilities and the game of chance. It seems that reversibility and irreversibility are opposite sides of the same coin. As physicists have already found through quantum mechanics, the full structure of the world is richer than our language can express and our brains comprehend. Many deep problems remain open for exploration, but at least we have made a start.

Further reading

PETER COVENEY and ROGER HIGHFIELD, *The Arrow of Time*, W. H. Allen, 1990.

18

Is the Universe a machine?

PAUL DAVIES

Chaos seems to provide a bridge between the deterministic laws of physics and the laws of chance, implying that the Universe is genuinely creative.

All science is founded on the assumption that the physical world is ordered. The most powerful expression of this order is found in the laws of physics. Nobody knows where these laws come from, nor why they apparently operate universally and unfailingly, but we see them at work all around us: in the rhythm of night and day, the pattern of planetary motions, the regular ticking of a clock.

The ordered dependability of nature is not, however, ubiquitous. The vagaries of the weather, the devastation of an earthquake, or the fall of a meteorite seem to be arbitrary and fortuitous. Small wonder that our ancestors attributed these events to the moodiness of the gods. But how are we to reconcile these apparently random 'acts of God' with the supposed underlying lawfulness of the Universe?

The ancient Greek philosphers regarded the world as a battleground between the forces of order, producing cosmos, and those of disorder, which led to chaos. They believed that random or disordering processes were negative, evil influences. Today, we don't regard the role of chance in nature as malicious, merely as blind. A chance event may act constructively, as in biological evolution, or destructively, such as when an aircraft fails from metal fatigue.

Though individual chance events may give the impression of lawlessness, disorderly processes, as a whole, may still display statistical regularities. Indeed, casino managers put as much faith in the laws of chance as engineers put in the laws of physics. But this raises something of a paradox. How can the same physical processes obey both the laws of physics and the laws of chance?

Following the formulation of the laws of mechanics by Isaac Newton in the 17th century, scientists became accustomed to thinking of the Universe as a gigantic mechanism. The most extreme form of this doctrine was strikingly expounded by Pierre Simon de Laplace in the 19th century. He envisaged every particle of matter as unswervingly locked in the embrace of strict mathematical laws of motion. These laws dictated the behaviour of even the smallest atom in the most minute detail. Laplace argued that, given the state of the Universe at any one instant, the entire cosmic future would be uniquely fixed, to infinite precision, by Newton's laws.

The concept of the Universe as a strictly deterministic machine governed by eternal laws profoundly influenced the scientific world view, standing as it did in stark contrast to the old Aristotelian picture of the cosmos as a living organism. A machine can have no 'free will'; its future is rigidly determined from the beginning of time. Indeed, time ceases to have much physical significance in this picture, for the future is already contained in the present. Time merely turns the pages of a cosmic history book that is already written.

Implicit in this somewhat bleak mechanistic picture was the belief that there are actually no truly chance processes in nature. Events may appear to us to be random but, it was reasoned, this could be attributed to human ignorance about the details of the processes concerned. Take, for example, Brownian motion. A tiny particle suspended in a fluid can be observed to execute a haphazard zigzag movement as a result of the slightly uneven buffeting it suffers at the hands of the fluid molecules that bombard it. Brownian motion is the archetypical random, unpredictable process. Yet, so the argument ran, if we could follow in detail the activities of all the individual molecules involved, Brownian motion would be every bit as predictable and deterministic as clockwork. The apparently random motion of the

Brownian particle is attributed solely to the lack of information about the myriads of participating molecules, arising from the fact that our senses are too coarse to permit detailed observation at the molecular level.

For a while, it was commonly believed that apparently 'chance' events were always the result of our ignoring, or effectively averaging over, vast numbers of hidden variables, or degrees of freedom. The toss of a coin or a die, the spin of a roulette wheel – these would no longer appear random if we could observe the world at the molecular level. The slavish conformity of the cosmic machine ensured that lawfulness was folded up in even the most haphazard events, albeit in an awesomely convoluted tangle.

Two major developments of the 20th century have, however, put paid to the idea of a clockwork universe. First there was quantum mechanics. At the heart of quantum physics lies Heisenberg's uncertainty principle, which states that everything we can measure is subject to truly random fluctuations. Quantum fluctuations are not the result of human limitations or hidden degrees of freedom; they are inherent in the workings of nature on an atomic scale. For example, the exact moment of decay of a particular radioactive nucleus is intrinsically uncertain. An element of genuine unpredictability is thus injected into nature.

Despite the uncertainty principle, there remains a sense in which quantum mechanics is still a deterministic theory. Although the outcome of a particular quantum process might be undetermined, the relative probabilities of different outcomes evolve in a deterministic manner. What this means is that you cannot know in any particular case what will be the outcome of the 'throw of the quantum dice', but you can know completely accurately how the betting odds vary from moment to moment. As a statistical theory, quantum mechanics remains deterministic. Quantum physics thus builds chance into the very fabric of reality, but a vestige of the Newtonian-Laplacian world view remains.

Then along came chaos. As previous chapters have discussed, the essential ideas of chaos were already present in the work of the mathematician Henri Poincaré at the turn of the century, but it is only in recent years, especially with the advent of fast electronic computers, that people have appreciated the full significance of chaos theory.

The key feature of a chaotic process concerns the way that predictive errors evolve with time. Let me first give an example of a non-chaotic system: the motion of a simple pendulum. Imagine two identical pendulums swinging in exact synchronism. Suppose that one pendulum is slightly disturbed so that its motion gets a little out of step with the other pendulum. This discrepancy, or phase shift, remains small as the pendulums go on swinging.

Faced with the task of predicting the motion of a simple pendulum, one could measure the position and velocity of the bob at some instant, and use Newton's laws to compute the subsequent behaviour. Any error in the initial measurement propagates through the calculation and appears as an error in the prediction. For the simple pendulum, a small input error implies a small output error in the predictive computation. In a typical non-chaotic system, errors accumulate with time. Crucially, though, the errors grow only in proportion to the time (or perhaps a small power thereof), so they remain relatively manageable.

Now let me contrast this property with that of a chaotic system. Here a small starting difference between two identical systems will rapidly grow. In fact, the hallmark of chaos is that the motions diverge exponentially fast. Translated into a prediction problem, this means that any input error multiplies itself at an escalating rate as a function of prediction time, so that before long it engulfs the calculation, and all predictive power is lost. Small input errors thus swell to calculation-wrecking size in very short order.

The distinction between chaotic and non-chaotic behaviour is well illustrated by the case of the spherical pendulum, this being a pendulum free to swing in two directions, as described by David Tritton in Chapter 2. In practice, this could be a ball suspended on the end of a string. If the system is driven in a plane by a periodic motion applied at the pivot, it will start to swing about. After a while, it may settle into a stable and entirely predictable pattern of motion, in which the bob traces out an elliptical path with the driving frequency. However, if you alter the driving frequency slightly, this regular motion may give way to chaos, with the bob swinging first this way and then that, doing a few clockwise turns, then a few anticlockwise turns in an apparently random manner.

The randomness of this system does not arise from the effect of myriads of hidden degrees of freedom. Indeed, by modelling mathematically only the three observed degrees of freedom (the three possible directions of motion), one may show that the behaviour of the pendulum is nonetheless random. And this is in spite of the fact that the mathematical model concerned is strictly deterministic.

It used to be supposed that determination went hand in hand with predictability, but we can now see that this need not be the case. A deterministic system is one in which future states are completely determined, through some dynamical law, by preceding states. There is thus a one-to-one association between earlier and later states. In computational terms, this suggests a one-to-one association between the input and the output of a predictive calculation. But now we must remember that any predictive computation will contain some input errors because we cannot measure physical quantities to unlimited precision. Moreover, computers can handle only finite quantities of data anyway.

The situation is represented geometrically in Figure 18.1. The fan of straight lines establishes a one-to-one correspondence

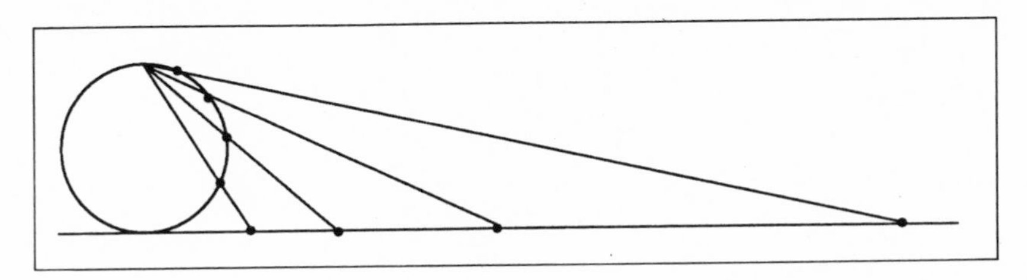

Figure 18.1

between points on the arc of the circle and points on the horizontal line. In the idealized case of perfect geometrical forms consisting of infinitesimally thin continuous lines and points of zero size, this correspondence is meaningful. But no real geometrical forms can be like this. As the top of the circle is approached, so points from a smaller and smaller arc are associated with a bigger and bigger segment of the horizontal line. (Think of points near the top of the arc as analogous to the initial conditions of a chaotic system, and points towards the right of the horizontal line as predicted values at later and later times.) The slightest uncertainty about one's position on the arc leads to a huge uncertainty about the corresponding point on the line segment. The one-to-one association becomes smudged into meaninglessness.

We might call this the fiction of the real line. The Ancient Greeks realized that points on a line could be labelled by numbers according to their distance from one end. Figure 18.2 shows a segment from 0 to 1. Fractions, such as 2/3 and 137/554, could be used to label the points in between. The Greeks called these numbers 'rational' (as in ratio). By using enough digits in the numerators and denominators we can choose a fraction that marks a place arbitrarily close to any designated point on the line. Nevertheless, it is readily shown that continuous line segments cannot have all their points labelled this way. That requires not only all possible rational numbers, but all irrational numbers too. An irrational number cannot be expressed as one whole number divided by another. It may instead be expressed as a decimal, with an infinite number of digits.

The set of all rational and irrational numbers form what mathematicians call the real numbers, and they underlie almost all modern theories of physics. The very notion of continuous mechanical processes, epitomized by Newton's calculus which

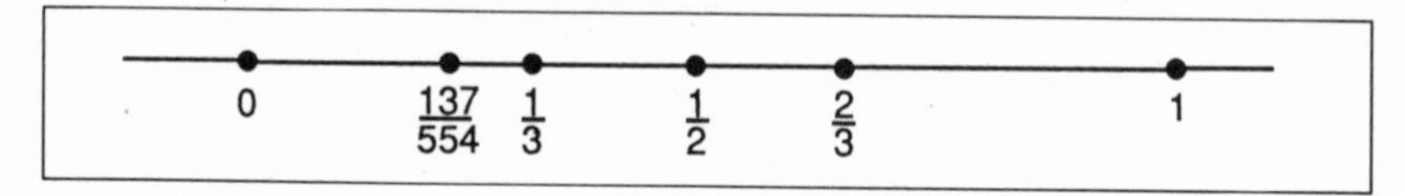

Figure 18.2

he formulated to describe them, is rooted in the concept of real numbers. Some real numbers, such as $1/2=0{\cdot}5$ or $1/3=0{\cdot}3333\ldots$, can be expressed compactly. But a typical real number has a decimal expansion consisting of an infinite string of digits with no systematic pattern to it, in other words, it is a random sequence. It follows that to specify such a number involves an infinite quantity of information. This is clearly impossible, even in principle. Even if we were to commandeer the entire observable Universe and employ it as a digital computer, its information storage capacity would be finite. Thus, the notion of a continuous line described by real numbers is exposed as a mathematical fiction.

Now consider the consequences for a chaotic system. Determinism implies predictability only in the idealized limit of infinite precision. In the case of the pendulum, for example, the behaviour will be determined uniquely by the initial conditions. The initial data includes the position of the bob, so exact predictability demands that we must assign the real number to the position that correctly describes the distance of the bob's centre from a fixed point. And this infinite precision is, as we have seen, impossible.

In a non-chaotic system this limitation is not so serious because the errors expand only slowly. But in a chaotic system errors grow at an accelerating rate. Suppose there is an uncertainty in, say, the fifth significant figure, and that this affects the prediction of how the system is behaving after a time t. A more accurate analysis might reduce the uncertainty to the tenth significant figure. But the exponential nature of error growth implies that the uncertainty now manifests itself after a time $2t$. So a hundred-thousand-fold improvement in initial accuracy achieves a mere doubling of the predictability span. It is this 'sensitivity to initial conditions' that leads to well-known statements about the flapping of butterflies' wings in the Amazonian jungle causing a tornado in Texas.

Chaos evidently provides us with a bridge between the laws of physics and the laws of chance. In a sense, chance or random events can indeed always be traced to ignorance about details, but whereas Brownian motion appears random because of the enormous number of degrees of freedom we are voluntarily overlooking, deterministic chaos appears random because we are necessarily ignorant of the ultra-fine detail of just a few degrees of freedom. And whereas Brownian chaos is complicated because the molecular bombardment is itself a complicated process, the motion of, say, the spherical pendulum is complicated even though the system itself is very simple. Thus, complicated behaviour does not necessarily imply complicated forces or laws. So the study of chaos has revealed how it is possible to reconcile the complexity of a physical world displaying haphazard and capricious behaviour with the order and simplicity of underlying laws of nature.

Though the existence of deterministic chaos comes as a surprise, we should not forget that nature is not, in fact, deterministic anyway. The indeterminism associated with quantum effects will intrude into the dynamics of all systems, chaotic or otherwise, at the atomic level. It might be supposed that quantum uncertainty would combine with chaos to amplify the unpredictability of the Universe. Curiously, however, quantum mechanics seems to have a subduing effect on chaos, as described by Michael Berry in Chapter 15. A number of model systems that are chaotic at the classical level are found to be non-chaotic when quantized. At this stage, the experts are divided about whether quantum chaos is possible, or how it would show itself if it did exist. Though the topic will undoubtedly prove important for atomic and molecular physics, it is of little relevance to the behaviour of macroscopic objects, or to the Universe as a whole.

What can we conclude about Laplace's image of a clockwork universe? The physical world contains a wide range of both chaotic and non-chaotic systems. Those that are chaotic have severely limited predictability, and even one such system would rapidly exhaust the entire Universe's capacity to compute its behaviour. It seems, then, that the Universe is incapable of digitally computing the future behaviour of even a small part of itself, let alone all of itself. Expressed more dramatically, the Universe is its own fastest simulator.

This conclusion is surely profound. It means that, even accepting a strictly deterministic account of nature, the future states of the Universe are in some sense 'open'. Some people have seized on this openness to argue for the reality of human free will. Others claim that it bestows upon nature an element of creativity, an ability to bring forth that which is genuinely new, something not already implicit in earlier states of the Universe, save in the idealized fiction of the real numbers. Whatever the merits of such sweeping claims, it seems safe to conclude from the study of chaos that the future of the Universe is not irredeemably fixed. The final chapter of the great cosmic book has yet to be written.

Further reading

PAUL DAVIES, *The Cosmic Blueprint*, Heinemann, 1987.

Notes on contributors

IAN PERCIVAL is professor of applied mathematics at Queen Mary and Westfield College, London. He has been head of a group studying chaos since 1972.

DAVID TRITTON is at the department of physics at the University of Newcastle upon Tyne.

FRANCO VIVALDI is in the School of Mathematical Sciences at Queen Mary and Westfield College, London.

IAN STEWART is a mathematician in the University of Warwick's Nonlinear Systems Laboratory.

TOM MULLIN is leader of a research group studying nonlinear systems in the Clarendon Laboratory at Oxford.

TIM PALMER is head of the predictability and diagnostics section of the European Centre for Medium-Range Weather Forecasts in Reading, Berkshire.

ROBERT MAY is a Royal Society Research Professor in the zoology department at the University of Oxford and at Imperial College, London.

CARL MURRAY is a reader in mathematics and astronomy in the Astronomy Unit, Queen Mary and Westfield College, University of London.

STEPHEN SCOTT is a lecturer in physical chemistry at Leeds University. He does experimental, theoretical and computational studies of chemical chaos.

BENOIT MANDELBROT is a physicist at the Thomas J. Watson Research Center of IBM, Yorktown Heights, New York, and a mathematician at Yale University. He originated fractal geometry and has played a major role in developing it.

CAROLINE SERIES is a reader in mathematics at the University of Warwick.

ALLAN MCROBIE is Wolfson research fellow and J. M. T. THOMPSON is professor of civil engineering, both at University College London.

JIM LESURF is a lecturer in the department of physics and astronomy at the University of St Andrews.

ROBERT SAVIT is a professor of physics at the University of Michigan. His research is in theoretical physics as well as other applications of nonlinear dynamics.

MICHAEL BERRY is a professor of physics at Bristol University.

GREGORY CHAITIN is a member of the theoretical physics group at the Thomas J. Watson Research Center in Yorktown Heights, New York, part of IBM's research division.

PETER COVENEY was a lecturer in physical chemistry at the University of Wales, Bangor. He is now a programme leader at Schlumberger Cambridge Research in Cambridge.

PAUL DAVIES is professor of mathematical physics at the University of Adelaide.

Acknowledgements

Grateful acknowledgement is given to the following for the use of figures and plates.

Figure 4.4 Anne Skeldon/Clarendon Laboratory, Oxford. Figures 5.1 and 5.3–5.6. Tom Mullin. Figure 7.5 A. Shrier [chick embryos]; A. Shrier, M. Guevara and L. Glass [graphs]. Figure 8.3 C. Murray. Figures 12.2–12.5 Allan McRobie, UCL. Figure 15.5 D. Wintgen, H. Friedlich and O. Bohigas. Figure 15.6 A. M. Odlyzko. Plates 1 and 2 Mario Markus and Benno Hess/Dortmund, Germany. Plate 3 Stuart Bebb and Tim Price/Clarendon Laboratory, Oxford. Plates 4–9 David Pottinger/IBM UK Scientific Centre, Winchester. Plates 10 and 11 Tom Mullin. Plate 12 James Yorke/University of Maryland. Plate 13 Stefan Müller, Theo Plesser and Benno Hess/Dortmund, Germany. Plate 14 University of Leeds. Plate 15 Heike Schuster and Martin Gerhardt/Dortmund, Germany. Plates 16 and 19 Richard Voss. Plate 17 F. Kenton Musgrave and B. A. Mandelbrot. Plate 18 B. A. Mandelbrot. Plate 20 A. Norton. Plate 21 Roman Tomaschitz. Plate 22 C. Leger. Plates 23–25 Allan McRobie. Plates 26 and 27 Jim Allan/University of St Andrews.